ELEMENTS

OF

GEOMETRY

AND

CONIC SECTIONS.

BY ELIAS LOOMIS, LL.D.,

PROFESSOR OF NATURAL PHILOSOPHY AND ASTRONOMY IN YALE COLLEGE, AND AUTHOR OF A "COURSE OF MATHEMATICS."

FIFTEENTH EDITION.

NEW YORK:

HARPER & BROTHERS, PUBLISHERS,

329 & 331 PEARL STREET,

(FRANKLIN SQUARE)

1861.

TO THE

HON THEODORE FRELINGHUYSEN, LL.D,

CHANCELLOR OF THE UNIVERSITY OF THE CITY OF NEW YORK,

THE FRIEND OF EDUCATION, THE PATRIOT STATESMAN,

AND THE CHRISTIAN PHILANTHROPIST,

This Work

IS RESPECTFULLY DEDICATED

BY

THE AUTHOR

PREFACE.

In the following treatise, an attempt has been made to combine the peculiar excellencies of Euclid and Legendre. The Elements of Euclid have long been celebrated as furnishing the most finished specimens of logic; and on this account they still retain their place in many seminaries of education, notwithstanding the advances which science has made in modern times. But the deficiencies of Euclid, particularly in Solid Geometry, are now so palpable, that few institutions are content with a simple translation from the original Greek. The edition of Euclid chiefly used in this country, is that of Professor Playfair, who has sought, by additions and supplements, to accommodate the Elements of Euclid to the present state of the mathematical sciences. But, even with these additions, the work is incomplete on Solids, and is very deficient on Spherical Geometry. Moreover, the additions are often incongruous with the original text; so that most of those who adhere to the use of Playfair's Euclid, will admit that something is still wanting to a perfect treatise. At most of our colleges, the work of Euclid has been superseded by that of Legendre. It seems superfluous to undertake a defense of Legendre's Geometry, when its merits are so generally appreciated. No one can doubt that, in respect of comprehensiveness and scientific arrangement, it is a great improvement upon the Elements of Euclid. Nevertheless, it should ever be borne in mind that, with most students in our colleges, the ultimate object is not to make profound mathematicians, but to make good reasoners on ordinary subjects. In order to secure this advantage, the learner should be trained, not merely to give the outline of a demonstration, but to state every part of the argument with minuteness and in its natural order. Now, although the model of Legendre is, for the most part, excellent, his demonstrations are often mere skeletons. They contain, indeed, the essential part of an argument; but the general student does not derive from them the highest benefit which may accrue from the study of Geometry as an exercise in reasoning.

While, then, in the following treatise, I have, for the most part, followed the arrangement of Legendre, I have aimed to give his demonstrations somewhat more of the logical method of Euclid. I have also made some changes in arrangement. Several of Legendre's propositions have been degraded to the rank of corollaries, while some of his corollaries or scholiums have been elevated to the dignity of primary propositions. His lemmas have been proscribed entirely, and most of his scholiums have received the more appropriate title of corollary. The quadrature of the circle is developed in an order somewhat different from any thing I have elsewhere seen. The propositions are all enunciated in general terms, with the utmost brevity which is consistent with clearness; and, in order to remind the student to conclude his recitation with the enunciation of the proposition, the leading words are repeated at the close of each demonstration. As the time given to mathematics in our colleges is limited, and a variety of subjects demand attention, no attempt has been made to render this a *complete* record of all the known propositions of Geometry. On the contrary, nearly every thing has been excluded which is not essential to the student's progress through the subsequent parts of his mathematical course.

Considerable attention has been given to the construction of the diagrams. I have aimed to reduce them all to nearly uniform dimensions, and to make them tolerable approximations to the objects they were designed to represent. I have made free use of dotted lines. Generally, the black lines are used to represent those parts of a figure which are directly involved in the statement of the proposition; while the dotted lines exhibit the parts which are added for the purposes of demonstration. In Solid Geometry the dotted lines commonly denote the parts which would be concealed by an opaque solid; while in a few cases, for peculiar reasons, both of these rules have been departed from. Throughout Solid Geometry the figures have generally been shaded, which addition, it is hoped, will obviate some of the difficulties of which students frequently complain.

The short treatise on the Conic Sections appended to this volume is designed particularly for those who have not time or inclination for the study of Analytical Geometry. Some acquaintance with the properties of the Ellipse and Parabola is indispensable as a preparation for the study of Mechanics and Astronomy. Those who pursue the study of Analytical Geometry can omit this treatise on the Conic Sections if it should be thought desirable. It is believed, however, that some knowledge of the properties of these curves, derived from geometrical methods, forms an excellent preparation for the Algebraical and more general processes of Analytical Geometry.

CONTENTS.

PLANE GEOMETRY.

SOLID GEOMETRY.

CONIC SECTIONS.

N.B.—When reference is made to a Proposition in the same Book, only the number of the Proposition is given; but when the Proposition is found in a different Book, the number of the Book is also specified.

ELEMENTS OF GEOMETRY.

BOOK I.

GENERAL PRINCIPLES.

Definitions.

1. GEOMETRY is that branch of Mathematics which treats of the properties of extension and figure.

Extension has three dimensions, length, breadth, and thickness.

2. A *line* is that which has length, without breadth or thickness.

The extremities of a line are called *points*. A point, therefore, has position, but not magnitude.

3. A *straight line* is the shortest path from one point to another.

4. Every line which is neither a straight line, nor composed of straight lines, is a *curved* line.

Thus, AB is a straight line, ACDB is a *broken* line, or one composed of straight lines, and AEB is a curved line.

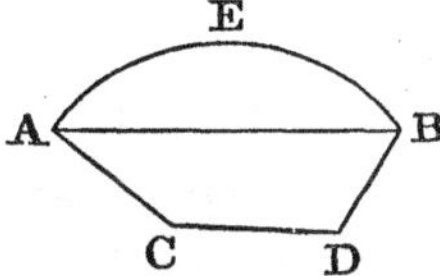

5. A *surface* is that which has length and breadth, without thickness.

6. A *plane* is a surface in which any two points being taken, the straight line which joins them lies wholly in that surface.

7. Every surface which is neither a plane, nor composed of plane surfaces, is a *curved surface.*

8. A *solid* is that which has length, breadth, and thickness, and therefore combines the three dimensions of extension.

9. When two straight lines meet together, their inclina-

tion, or opening, is called an *angle.* The point of meeting is called the *vertex*, and the lines are called the *sides* of the angle.

If there is only one angle at a point, it may be denoted by a letter placed at the vertex, as the angle at A.

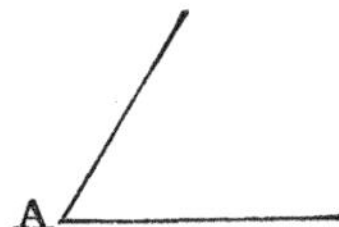

But if several angles are at one point, any one of them is expressed by three letters, of which the middle one is the letter at the vertex.

Thus, the angle which is contained by the straight lines BC, CD, is called the angle BCD, or DCB.

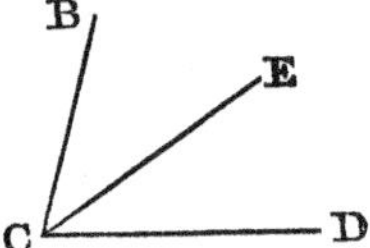

Angles, like other quantities, may be added, subtracted, multiplied, or divided. Thus, the angle BCD is the sum of the two angles BCE, ECD; and the angle ECD is the difference between the two angles BCD, BCE.

10. When a straight line, meeting another straight line makes the adjacent angles equal to one another, each of them is called a *right angle,* and the straight line which meets the other is called a *perpendicular* to it.

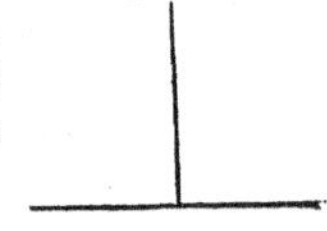

11. An *acute angle* is one which is less than a right angle.

An *obtuse angle* is one which is greater than a right angle.

12. *Parallel* straight lines are such as are in the same plane, and which, being produced ever so far both ways, do not meet.

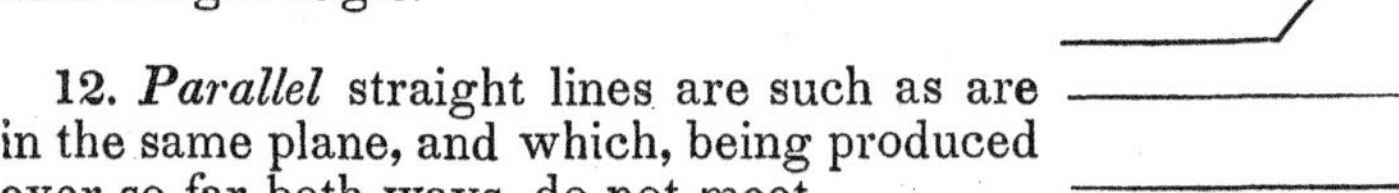

13. A *plane figure* is a plane terminated on all sides by lines either straight or curved.

If the lines are straight, the space they inclose is called a *rectilineal figure,* or *polygon,* and the lines themselves, taken together, form the *perimeter* of the polygon.

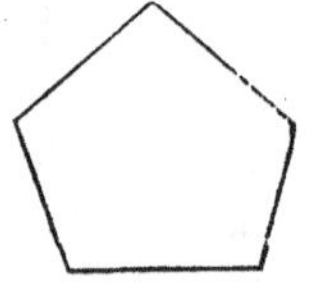

14. The polygon of three sides is the simplest of all, and is called a *triangle;* that of four sides is called a *quadrilateral:* that of five, a *pentagon;* that of six, a *hexagon,* &c.

15. An *equilateral* triangle is one which has its three sides equal.

An *isosceles* triangle is that which has only two sides equal.

A *scalene* triangle is one which has three unequal sides.

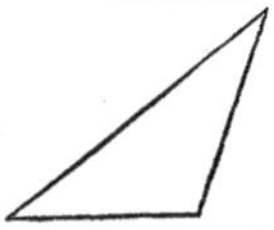

16. A *right-angled* triangle is one which has a right angle. The side opposite the right angle is called the *hypothenuse.*

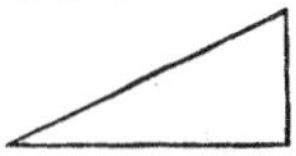

An *obtuse-angled* triangle is one which has an obtuse angle. An *acute-angled* triangle is one which has three acute angles.

17. Of quadrilaterals, a *square* is that which has all its sides equal, and its angles right angles.

A *rectangle* is that which has all its angles right angles, but all its sides are not necessarily equal.

A *rhombus* is that which has all its sides equal, but its angles are not right angles.

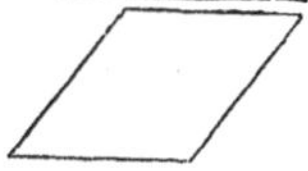

A *parallelogram* is that which has its opposite sides parallel.

A *trapezoid* is that which has only two sides parallel.

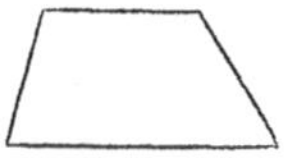

18. The *diagonal* of a figure is a line which joins the vertices of two angles not adjacent to each other.

Thus, AC, AD, AE are diagonals.

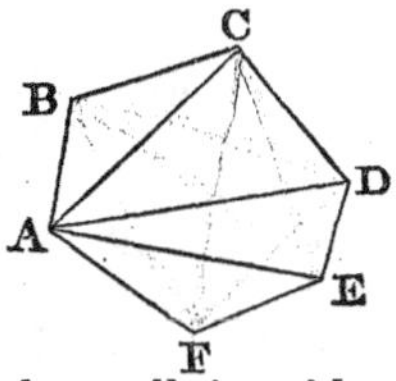

19. An *equilateral* polygon is one which has all its sides equal. An *equiangular* polygon is one which has all its angles equal.

20. Two polygons are *mutually equilateral* when they have all the sides of the one equal to the corresponding sides of the other, each to each, and arranged in the same order.

Two polygons are *mutually equiangular* when they have

all the angles of the one equal to the corresponding angles of the other, each to each, and arranged in the same order.

In both cases, the equal sides, or the equal angles, are called *homologous* sides or angles.

21. An *axiom* is a self-evident truth.

22. A *theorem* is a truth which becomes evident by a train of reasoning called a *demonstration.*

A *direct* demonstration proceeds from the premises by a regular deduction.

An *indirect* demonstration shows that any supposition contrary to the truth advanced, necessarily leads to an absurdity.

23. A *problem* is a question proposed which requires a *solution.*

24. A *postulate* requires us to admit the possibility of an operation.

25. A *proposition* is a general term for either a theorem, or a problem.

One proposition is the *converse* of another, when the conclusion of the first is made the supposition in the second.

26. A *corollary* is an obvious consequence, resulting from one or more propositions.

27. A *scholium* is a remark appended to a proposition.

28. An *hypothesis* is a supposition made either in the enunciation of a proposition, or in the course of a demonstration.

Axioms.

1. Things which are equal to the same thing are equal to each other.
2. If equals are added to equals, the wholes are equal.
3. If equals are taken from equals, the remainders are equal.
4. If equals are added to unequals, the wholes are unequal.
5. If equals are taken from unequals, the remainders are unequal.
6. Things which are doubles of the same thing are equal to each other.
7. Things which are halves of the same thing are equal to each other.
8. Magnitudes which coincide with each other, that is, which exactly fill the same space, are equal.
9. The whole is greater than any of its parts.
10. The whole is equal to the sum of all its parts.
11. From one point to another only one straight line can be drawn.

12. Two straight lines, which intersect one another, can not both be parallel to the same straight line.

Explanation of Signs.

For the sake of brevity, it is convenient to employ, to some extent, the signs of Algebra in Geometry. Those chiefly employed are the following:

The sign = denotes that the quantities between which it stands are equal; thus, the expression A=B signifies that A is equal to B.

The sign + is called *plus*, and indicates addition; thus A+B represents the sum of the quantities A and B.

The sign — is called *minus*, and indicates subtraction; thus, A—B represents what remains after subtracting B from A.

The sign × indicates multiplication; thus, A×B denotes the product of A by B. Instead of the sign ×, a point is sometimes employed; thus, A.B is the same as A×B. The same product is also sometimes represented without any intermediate sign, by AB; but this expression should not be employed when there is any danger of confounding it with the line AB.

A parenthesis ( ) indicates that several quantities are to be subjected to the same operation; thus, the expression A×(B+C—D) represents the product of A by the quantity B+C—D.

The expression $\frac{A}{B}$ indicates the quotient arising from dividing A by B.

A number placed before a line or a quantity is to be regarded as a multiplier of that line or quantity; thus, 3AB denotes that the line AB is taken three times; ½A denotes the half of A.

The square of the line AB is denoted by AB^2; its cube by AB^3.

The sign $\sqrt{\ }$ indicates a root to be extracted; thus, $\sqrt{2}$ denotes the square root of 2; $\sqrt{A \times B}$ denotes the square root of the product of A and B.

N.B.—*The first six books treat only of plane figures, or figures drawn on a plane surface.*

PROPOSITION I. THEOREM.

All right angles are equal to each other.

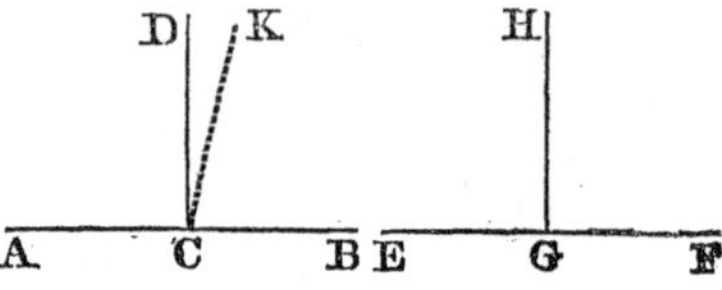

Let the straight line CD be perpendicular to AB, and GH to EF; then, by definition 10, each of the angles ACD, BCD, EGH, FGH, will be a right angle; and it is to be proved that the angle ACD is equal to the angle EGH.

Take the four straight lines AC, CB, EG, GF, all equal to each other; then will the line AB be equal to the line EF (Axiom 2). Let the line EF be applied to the line AB, so that the point E may be on A, and the point F on B; then will the lines EF, AB coincide throughout; for otherwise two different straight lines might be drawn from one point to another, which is impossible (Axiom 11). Moreover, since the line EG is equal to the line AC, the point G will fall on the point C; and the line EG, coinciding with AC, the line GH will coincide with CD. For, if it could have any other position, as CK, then, because the angle EGH is equal to FGH (Def. 10), the angle ACK must be equal to BCK, and therefore the angle ACD is less than BCK. But BCK is less than BCD (Axiom 9); much more, then, is ACD less than BCD, which is impossible, because the angle ACD is equal to the angle BCD (Def. 10); therefore, GH can not but coincide with CD, and the angle EGH coincides with the angle ACD, and is equal to it (Axiom 8). Therefore, all right angles are equal to each other.

PROPOSITION II. THEOREM.

The angles which one straight line makes with another, upon one side of it, are either two right angles, or are together equal to two right angles.

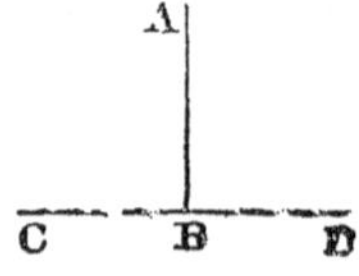

Let the straight line AB make with CD, upon one side of it, the angles ABC, ABD; these are either two right angles, or are together equal to two right angles.

For if the angle ABC is equal to ABD, each of them is a right angle (Def. 10); but if not, suppose the line BE to be drawn from the point B, perpendicular to CD; then will each of the angles CBE, DBE be a right angle. Now the angle CBA is equal to the sum of the two angles CBE, EBA. To each of these equals add the angle ABD; then the sum of the two angles CBA, ABD will be equal to the sum of the three angles CBE, EBA, ABD (Axiom 2). Again, the angle DBE is equal to the sum of the two angles DBA, ABE. Add to each of these equals the angle EBC; then will the sum of the two angles DBE, EBC be equal to the sum of the three angles DBA, ABE, EBC. Now things that are equal to the same thing are equal to each other (Axiom 1); therefore, the sum of the angles CBA, ABD is equal to the sum of the angles CBE, EBD. But CBE, EBD are two right angles; therefore ABC, ABD are together equal to two right angles. Therefore, the angles which one straight line, &c.

Corollary 1. If one of the angles ABC, ABD is a right angle, the other is also a right angle.

Cor. 2. If the line DE is perpendicular to AB, conversely, AB will be perpendicular to DE.

For, because DE is perpendicular to AB, the angle DCA must be equal to its adjacent angle DCB (Def. 10), and each of them must be a right angle. But since ACD is a right angle, its adjacent angle, ACE, must also be a right angle (Cor. 1). Hence the angle ACE is equal to the angle ACD (Prop. I.), and AB is perpendicular to DE.

Cor. 3. The sum of all the angles BAC, CAD, DAE, EAF, formed on the same side of the line BF, is equal to two right angles; for their sum is equal to that of the two adjacent angles BAD, DAF.

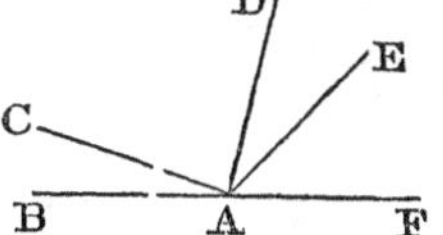

PROPOSITION III. THEOREM (*Converse of Prop. II.*).

If, at a point in a straight line, two other straight lines, upon the opposite sides of it, make the adjacent angles together equal to two right angles, these two straight lines are in one and the same straight line.

At the point B, in the straight line AB, let the two straight lines BC, BD, upon the opposite sides of AB, make the adjacent angles, ABC, ABD, together equal to two right angles·

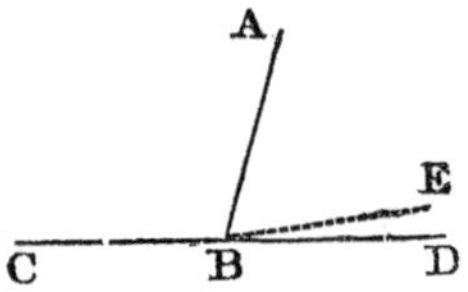

then will BD be in the same straight line with CB.

For, if BD is not in the same straight line with CB, let BE be in the same straight line with it; then, because the straight line CBE is met by the straight line AB, the angles ABC, ABE are together equal to two right angles (Prop. II.). But, by hypothesis, the angles ABC, ABD are together equal to two right angles; therefore, the sum of the angles ABC, ABE is equal to the sum of the angles ABC, ABD. Take away the common angle ABC, and the remaining angle ABE, is equal (Axiom 3) to the remaining angle ABD, the less to the greater, which is impossible. Hence BE is not in the same straight line with BC; and in like manner, it may be proved that no other can be in the same straight line with it but BD. Therefore, if at a point, &c.

PROPOSITION IV. THEOREM.

Two straight lines, which have two points common, coincide with each other throughout their whole extent, and form but one and the same straight line.

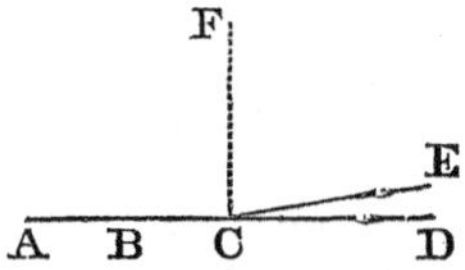

Let there be two straight lines, having the points A and B in common; these lines will coincide throughout their whole extent.

It is plain that the two lines must coincide between A and B, for otherwise there would be two straight lines between A and B, which is impossible (Axiom 11). Suppose, however, that, on being produced, these lines begin to diverge at the point C, one taking the direction CD, and the other CE. From the point C draw the line CF at rignt angles with AC; then, since ACD is a straight line, the angle FCD is a right angle (Prop. II, Cor. 1); and since ACE is a straight line, the angle FCE is also a right angle; therefore (Prop. I.), the angle FCE is equal to the angle FCD, the less to the greater, which is absurd. Therefore, two straight lines which have, &c.

PROPOSITION V. THEOREM.

If two straight lines cut one another, the vertical or opposite angles are equal.

Let the two straigh. lines, AB, CD, cut one another in the

point E; then will the angle AEC be equal to the angle BED, and the angle AED to the angle CEB.

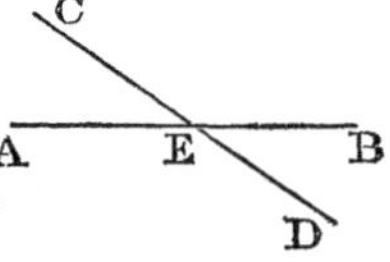

For the angles AEC, AED, which the straight line AE makes with the straight line CD, are together equal to two right angles (Prop. II.); and the angles AED, DEB, which the straight line DE makes with the straight line AB, are also together equal to two right angles; therefore, the sum of the two angles AEC, AED is equal to the sum of the two angles AED, DEB. Take away the common angle AED, and the remaining angle, AEC, is equal to the remaining angle DEB (Axiom 3). In the same manner, it may be proved that the angle AED is equal to the angle CEB. Therefore, if two straight lines, &c.

Cor. 1. Hence, if two straight lines cut one another, the four angles formed at the point of intersection, are together equal to four right angles.

Cor. 2. Hence, all the angles made by any number of straight lines meeting in one point, are together equal to four right angles.

PROPOSITION VI. THEOREM.

If two triangles have two sides, and the included angle of the one, equal to two sides and the included angle of the other, each to each, the two triangles will be equal, their third sides will be equal, and their other angles will be equal, each to each.

Let ABC, DEF be two triangles, having the side AB equal to DE, and AC to DF, and also the angle A equal to the angle D; then will the triangle ABC be equal to the triangle DEF.

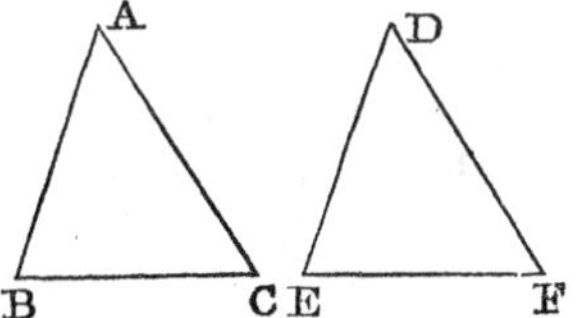

For, if the triangle ABC is applied to the triangle DEF, so that the point A may be on D, and the straight line AB upon DE, the point B will coincide with the point E, because AB is equal to DE; and AB, coinciding with DE, AC will coincide with DF, because the angle A is equal to the angle D. Hence, also, the point C will coincide with the point F, because AC is equal to DF. But the point B coincides with the point E; therefore the base BC will coincide with the base EF (Axiom 11), and will be equal to it. Hence, also, the whole triangle ABC will coincide with the whole triangle DEF, and will be equal to it and the remaining angles of the one, will coincide with the remaining angles of the other, and be equal to them, viz.: the angle ABC to the angle DEF, and the angle ACB to the angle DFE. Therefore, if two triangles, &c.

PROPOSITION VII. THEOREM.

If two triangles have two angles, and the included side of the one, equal to two angles and the included side of the other, each to each, the two triangles will be equal, the other sides will be equal, each to each, and the third angle of the one to the third angle of the other.

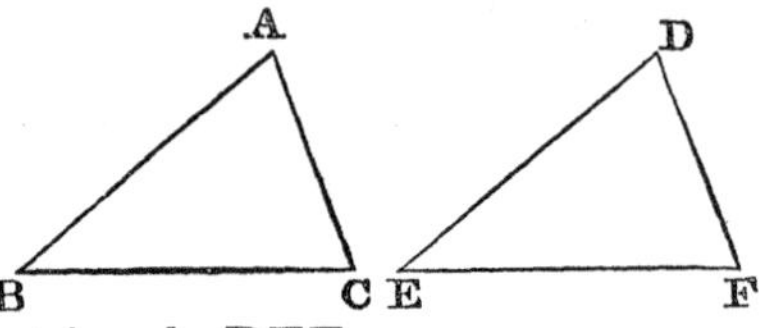

Let ABC, DEF be two triangles having the angle B equal to E, the angle C equal to F, and the included sides BC, EF equal to each other; then will the triangle ABC be equal to the triangle DEF.

For, if the triangle ABC is applied to the triangle DEF, so that the point B may be on E, and the straight line BC upon EF, the point C will coincide with the point F, because BC is equal to EF. Also, since the angle B is equal to the angle E, the side BA will take the direction ED, and therefore the point A will be found somewhere in the line DE. And, because the angle C is equal to the angle F, the line CA will take the direction FD, and the point A will be found somewhere in the line DF; therefore, the point A, being found at the same time in the two straight lines DE, DF, must fall at their intersection, D. Hence the two triangles ABC, DEF coincide throughout, and are equal to each other; also, the two sides AB, AC are equal to the two sides DE, DF, each to each, and the angle A to the angle D. Therefore, if two triangles, &c.

PROPOSITION VIII. THEOREM.

Any side of a triangle is less than the sum of the other two

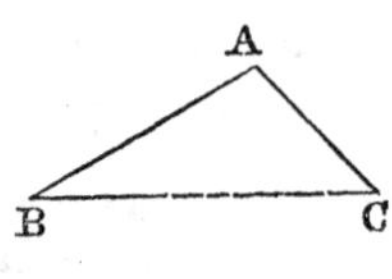

Let ABC be a triangle; any one of its sides is less than the sum of the other two, viz.: the side AB is less than the sum of AC and BC; BC is less than the sum of AB and AC; and AC is less than the sum of AB and BC.

For the straight line AB is the shortest path between the points A and B (Def. 3); hence AB is less than the sum of AC and BC. For the same reason, BC is less than the sum of AB and AC; and AC less than the sum of AB and BC Therefore, any two sides, &c.

PROPOSITION IX. THEOREM.

If, from a point within a triangle, two straight lines are drawn to the extremities of either side, their sum will be less than the sum of the other two sides of the triangle.

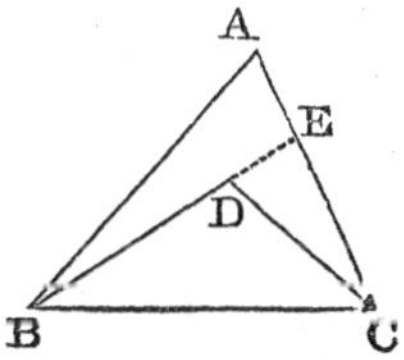

Let the two straight lines BD, CD be drawn from D, a point within the triangle ABC, to the extremities of the side BC; then will the sum of BD and DC be less than the sum of BA, AC, the other two sides of the triangle.

A E D B C

Produce BD until it meets the side AC in E; and, because one side of a triangle is less than the sum of the other two (Prop. VIII.), the side CD of the triangle CDE is less than the sum of CE and ED. To each of these add DB; then will the sum of CD and BD be less than the sum of CE and EB. Again, because the side BE of the triangle BAE is less than the sum of BA and AE, if EC be added to each, the sum of BE and EC will be less than the sum of BA and AC. But it has been proved that the sum of BD and DC is less than the sum of BE and EC; much more, then, is the sum of BD and DC less than the sum of BA and AC. Therefore, if from a point, &c.

PROPOSITION X. THEOREM.

The angles at the base of an isosceles triangle are equal to one another.

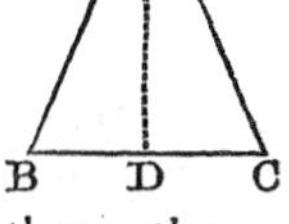

Let ABC be an isosceles triangle, of which the side AB is equal to AC; then will the angle B be equal to the angle C.

A B D C

For, conceive the angle BAC to be bisected by the straight line AD; then, in the two triangles ABD, ACD, two sides AB, AD, and the included angle in the one, are equal to the two sides AC, AD, and the included angle in the other; therefore (Prop. VI.), the angle B is equal to the angle C. Therefore, the angles at the base, &c.

Cor. 1. Hence, also, the line BD is equal to DC, and the angle ADB equal to ADC; consequently, each of these angles is a right angle (Def. 10). Therefore, *the line bisecting the vertical angle of an isosceles triangle bisects the base at right angles;* and, conversely, *the line bisecting the base of an isosceles triangle at right angles bisects also the vertical angle.*

Cor. 2. Every equilateral triangle is also equiangular.

Scholium. Any side of a triangle may be considered as its *base*, and the opposite angle as its *vertex;* but in an isosceles triangle, that side is usually regarded as the base, which is not equal to either of the others.

PROPOSITION XI. THEOREM (*Converse of Prop. X.*).

If two angles of a triangle are equal to one another, the opposite sides are also equal.

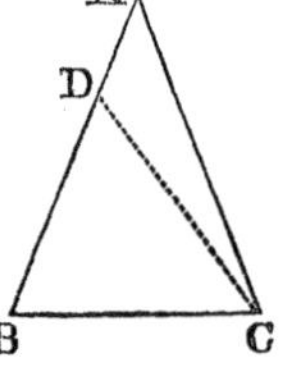

Let ABC be a triangle having the angle ABC equal to the angle ACB; then will the side AB be equal to the side AC.

For if AB is not equal to AC, one of them must be greater than the other. Let AB be the greater, and from it cut off DB equal to AC the less, and join CD. Then, because in the triangles DBC, ACB, DB is equal to AC, and BC is common to both triangles, also, by supposition, the angle DBC is equal to the angle ACB; therefore, the triangle DBC is equal to the triangle ACB (Prop. VI.), the less to the greater, which is absurd. Hence AB is not unequal to AC, that is, it is equal to it. Therefore, if two angles, &c.

Cor. Hence, every equiangular triangle is also equilateral.

PROPOSITION XII. THEOREM.

The greater side of every triangle is opposite to the greater angle; and, conversely, the greater angle is opposite to the greater side.

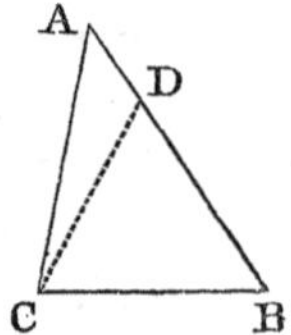

Let ABC be a triangle, having the angle ACB greater than the angle ABC; then will the side AB be greater than the side AC.

Draw the straight line CD, making the angle BCD equal to B; then, in the triangle CDB, the side CD must be equal to DB (Prop. XI.). Add AD to each, then will the sum of AD and DC be equal to the sum of AD and DB. But AC is less tnan the sum of AD and DC (Prop. VIII.); it is, therefore, less than AB.

Conversely, if the side AB is greater than the side AC, then will the angle ACB be greater than the angle ABC.

For if ACB is not greater than ABC, it must be either equal to it, or less. It is not equal, because then the side AB would be equal to the side AC (Prop. XI.), which is contrary to the supposition. Neither is it less, because then the side AB would be less than the side AC, according to the former part of this proposition; hence ACB must be greater than ABC. Therefore, the greater side, &c.

PROPOSITION XIII. THEOREM.

If two triangles have two sides of the one equal to two sides of the other, each to each, but the included angles unequal, the base of that which has the greater angle, will be greater than the base of the other.

Let ABC, DEF be two triangles, having two sides of the one equal to two sides of the other, viz.: AB equal to DE, and AC to DF, but the angle BAC greater than the angle EDF; then will the base BC be greater than the base EF.

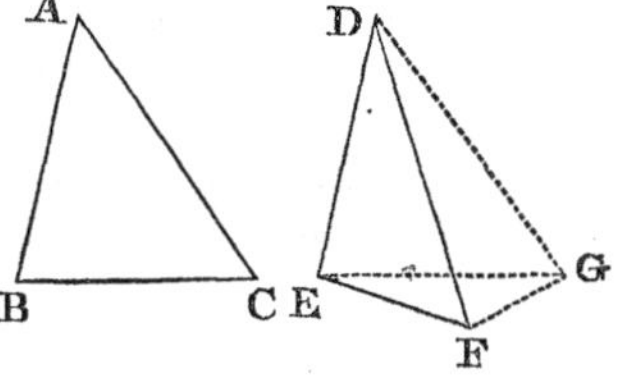

Of the two sides DE, DF, let DE be the side which is not greater than the other; and at the point D, in the straight line DE, make the angle EDG equal to BAC; make DG equal to AC or DF, and join EG, GF.

Because, in the triangles ABC, DEG, AB is equal to DE, and AC to DG; also, the angle BAC is equal to the angle EDG; therefore, the base BC is equal to the base EG (Prop. VI.). Also, because DG is equal to DF, the angle DFG is equal to the angle DGF (Prop. X.). But the angle DGF is greater than the angle EGF; therefore the angle DFG is greater than EGF; and much more is the angle EFG greater than the angle EGF. Now, in the triangle EFG, because the angle EFG is greater than EGF, and because the greater side is opposite the greater angle (Prop. XII.), the side EG is greater than the side EF. But EG has been proved equal to BC; and hence BC is greater than EF. Therefore, if two triangles, &c.

PROPOSITION XIV. THEOREM (*Converse of Prop XIII.*).

If two triangles have two sides of the one equal to two sides of the other, each to each, but the bases unequal, the angle contained by the sides of that which has the greater base, will be greater than the angle contained by the sides of the other.

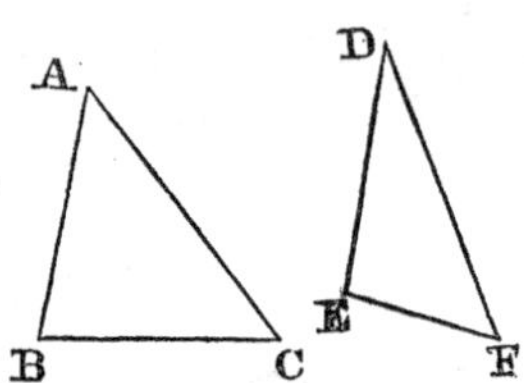

Let ABC, DEF be two triangles having two sides of the one equal to two sides of the other, viz.: AB equal to DE, and AC to DF, but the base BC greater than the base EF; then will the angle BAC be greater than the angle EDF.

For if it is not greater, it must be either equal to it, or less. But the angle BAC is not equal to the angle EDF, because then the base BC would be equal to the base EF (Prop. VI.), which is contrary to the supposition. Neither is it less, because then the base BC would be less than the base EF (Prop. XIII.), which is also contrary to the supposition; therefore, the angle BAC is not less than the angle EDF, and it has been proved that it is not equal to it; hence the angle BAC must be greater than the angle EDF. Therefore, if two triangles, &c.

PROPOSITION XV. THEOREM.

If two triangles have the three sides of the one equal to the three sides of the other, each to each, the three angles will also be equal, each to each, and the triangles themselves will be equal

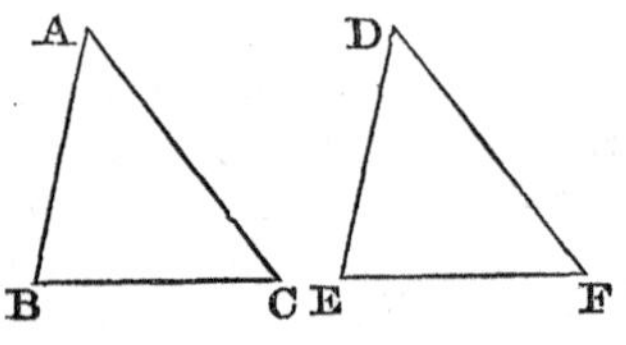

Let ABC, DEF be two triangles having the three sides of the one equal to the three sides of the other, viz.: AB equal to DE, BC to EF, and AC to DF; then will the three angles also be equal, viz.: the angle A to the angle D, the angle B to the angle E, and the angle C to the angle F.

For if the angle A is not equal to the angle D, it must be either greater or less. It is not greater, because then the base BC would be greater than the base EF (Prop. XIII.) which is contrary to the hypothesis; neither is it less, be cause then the base BC would be less than the base EF (Prop. XIII.), which is also contrary to the hypothesis. Therefore, the angle A must be equal to the angle D. In the same manner, it may be proved that the angle B is equal to the angle E, and the angle C to the angle F; hence the two triangles are equal. Therefore, if two triangles, &c.

Scholium. In equal triangles, the equal angles are opposite to the equal sides; thus, the equal angles A and D are opposite to the equal sides BC, EF.

PROPOSITION XVI. THEOREM.

From a point without a straight line, only one perpendicular can be drawn to that line.

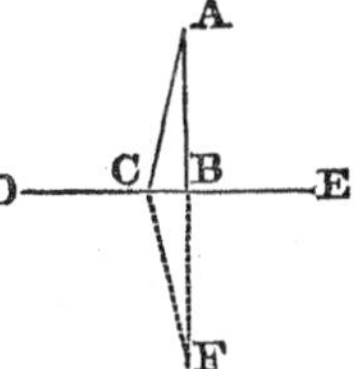

Let A be the given point, and DE the given straight line; from the point A only one perpendicular can be drawn to DE.

For, if possible, let there be drawn two perpendiculars AB, AC. Produce the line AB to F, making BF equal to AB, and join CF. Then, in the triangles ABC, FBC, because AB is equal to BF, BC is common to both triangles, and the angle ABC is equal to the angle FBC, being both right angles (Prop. II., Cor. 1); therefore, two sides and the included angle of one triangle, are equal to two sides and the included angle of the other triangle; hence the angle ACB is equal to the angle FCB (Prop. VI.). But, since the angle ACB is, by supposition, a right angle, FCB must also be a right angle; and the two adjacent angles BCA, BCF, being together equal to two right angles, the two straight lines AC, CF must form one and the same straight line (Prop. III.); that is, between the two points A and F, two straight lines, ABF, ACF, may be drawn, which is impossible (Axiom 11); hence AB and AC can not both be perpendicular to DE. Therefore, from a point, &c.

Cor. From the same point, C, in the line AB, more than one perpendicular to this line can not be drawn. For, if possible, let CD and CE be two perpendiculars; then, because CD is perpendicular to AB, the angle DCA is a right angle; and, because CE is perpendicular to AB, the angle ECA is also a right angle. Hence the angle ACD is equal to the angle ACE (Prop. I.), the less to the greater, which is absurd; therefore, CD and CE can not both be perpendicular to AB from the same point C.

PROPOSITION XVII. THEOREM.

If, from a point without a straight line, a perpendicular be drawn to this line, and oblique lines be drawn to different points:

1*st. The perpendicular will be shorter than any oblique line*

2*d. Two oblique lines, which meet the proposed line at equal distances from the perpendicular, will be equal.*

3*d. Of any two oblique lines, that which is further from the perpendicular will be the longer.*

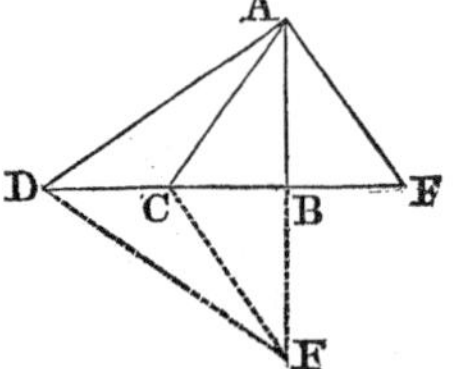

Let DE be the given straight line, and A any point without it. Draw AB perpendicular to DE; draw, also, the oblique lines AC, AD, AE. Produce the line AB to F, making BF equal to AB, and join CF, DF.

First. Because, in the triangles ABC, FBC, AB is equal to BF, BC is common to the two triangles, and the angle ABC is equal to the angle FBC, being both right angles (Prop. II., Cor. 1); therefore, two sides and the included angle of one triangle, are equal to two sides and the included angle of the other triangle; hence the side CF is equal to the side CA (Prop. VI.). But the straight line ABF is shorter than the broken line ACF (Prop. VIII.); hence AB, the half of ABF, is shorter than AC, the half of ACF. Therefore, the perpendicular AB is shorter than any oblique line, AC.

Secondly. Let AC and AE be two oblique lines which meet the line DE at equal distances from the perpendicular; they will be equal to each other. For, in the triangles ABC, ABE, BC is equal to BE, AB is common to the two triangles, and the angle ABC is equal to the angle ABE, being both right angles (Prop. I.); therefore, two sides and the included angle of one triangle are equal to two sides and the included angle of the other; hence the side AC is equal to the side AE (Prop. VI.). Wherefore, two oblique lines, equally distant from the perpendicular, are equal.

Thirdly. Let AC, AD be two oblique lines, of which AD is further from the perpendicular than AC; then will AD be longer than AC. For it has already been proved that AC is equal to CF; and in the same manner it may be proved that AD is equal to DF. Now, by Prop. IX., the sum of the two lines AC, CF is less than the sum of the two lines AD, DF. Therefore, AC, the half of ACF, is less than AD, the half of ADF; hence the oblique line which is furthest from the perpendicular is the longest. Therefore, if from a point, &c.

Cor. 1. The perpendicular measures the shortest distance of a point from a line, because it is shorter than any oblique line.

Cor. 2. It is impossible to draw three equal straight lines from the same point to a given straight line.

PROPOSITION XVIII. THEOREM.

If through the middle point of a straight line a perpendicular is drawn to this line:

1*st. Each point in the perpendicular is equally distant from the two extremities of the line.*

2*d. Any point out of the perpendicular is unequally distant from those extremities.*

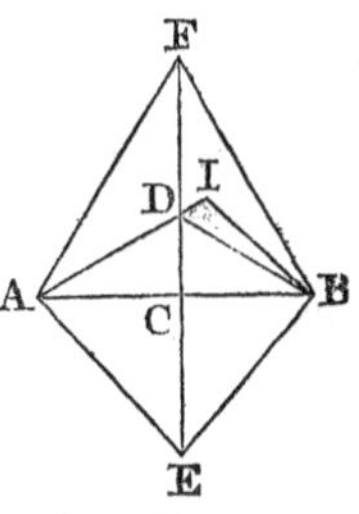

Let the straight line EF be drawn perpendicular to AB through its middle point, C.

First. Every point of EF is equally distant from the extremities of the line AB; for, since AC is equal to CB, the two oblique lines AD, DB are equally distant from the perpendicular, and are, therefore, equal (Prop. XVII.). So, also, the two oblique lines AE, EB are equal, and the oblique lines AF, FB are equal; therefore, every point of the perpendicular is equally distant from the extremities A and B.

Secondly. Let I be any point out of the perpendicular. Draw the straight lines IA, IB; one of these lines must cut the perpendicular in some point, as D. Join DB; then, by the first case, AD is equal to DB. To each of these equals add ID, then will IA be equal to the sum of ID and DB. Now, in the triangle IDB, IB is less than the sum of ID and DB (Prop. VIII.); it is, therefore, less than IA; hence, every point out of the perpendicular is unequally distant from the extremities A and B. Therefore, if through the middle point, &c.

Cor. If a straight line have two points, each of which is equally distant from the extremities of a second line, it will be perpendicular to the second line at its middle point.

PROPOSITION XIX. THEOREM.

If two right-angled triangles have the hypothenuse and a side of the one, equal to the hypothenuse and a side of the other each to each, the triangles are equal.

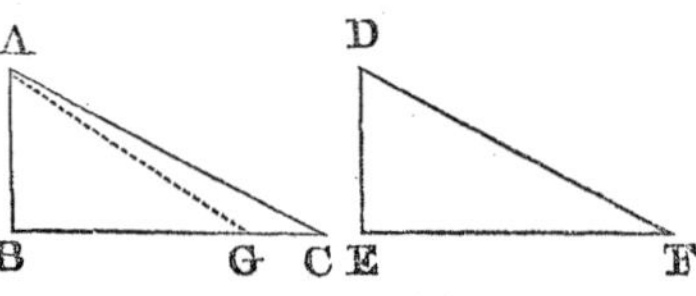

Let ABC, DEF be two right-angled triangles, having the hypothenuse AC and the side AB of the one, equal to the hypothenuse DF and side DE of the other; then will the side BC be equal to EF, and the triangle ABC to the triangle DEF.

For if BC is not equal to EF, one of them must be greater than the other. Let BC be the greater, and from it cut off BG equal to EF the less, and join AG. Then, in the triangles ABG, DEF, because AB is equal to DE, BG is equal to EF, and the angle B equal to the angle E, both of them being right angles, the two triangles are equal (Prop. VI.), and AG is equal to DF. But, by hypothesis, AC is equal to DF, and therefore AG is equal to AC. Now the oblique line AC, being further from the perpendicular than AG, is the longer (Prop. XVII.), and it has been proved to be equal, which is impossible. Hence BC is not unequal to EF, that is, it is equal to it; and the triangle ABC is equal to the triangle DEF (Prop. XV.). Therefore, if two right-angled triangles, &c

PROPOSITION XX. THEOREM.

Two straight lines perpendicular to a third line, are parallel.

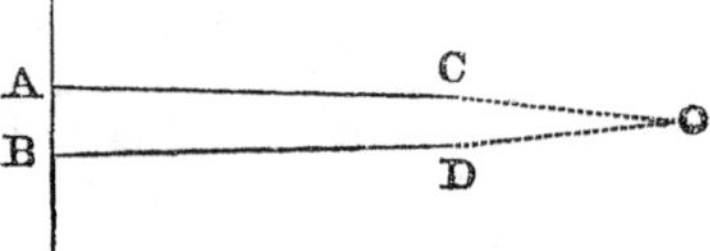

Let the two straight lines AC, BD be both perpendicular to AB; then is AC parallel to BD.

For if these lines are not parallel, being produced, they must meet on one side or the other of AB. Let them be produced, and meet in O; then there will be two perpendiculars, OA, OB, let fall from the same point, on the same straight line, which is impossible (Prop. XVI.). Therefore, two straight lines, &c.

PROPOSITION XXI. THEOREM.

If a straight line, meeting two other straight lines, makes the interior angles on the same side, together equal to two right angles, the two lines are parallel.

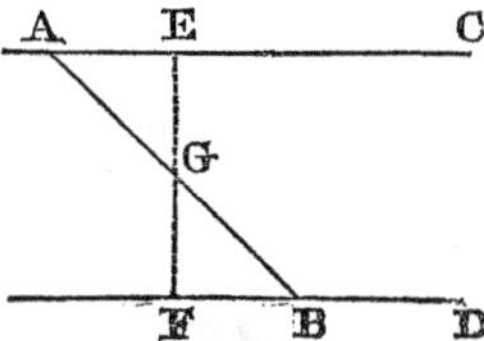

Let the straight line AB, which meets the two straight lines AC, BD, make the interior angles on the same side, BAC, ABD, together equal to two right angles; then is AC parallel to BD.

From G, the middle point of the line AB, draw EGF perpendicular to AC; it will also be perpendicular to BD. For the sum of the angles ABD and ABF is equal to two right angles (Prop. II.); and by hypothesis the sum of the angles ABD and BAC is equal to two right angles. Therefore, the sum of ABD and ABF is equal to the sum of ABD and BAC. Take away the common angle ABD, and the remainder, ABF, is equal to BAC; that is GBF is equal to GAE.

Again, the angle BGF is equal to the angle AGE (Prop V.); and, by construction, BG is equal to GA; hence the triangles BGF, AGE have two angles and the included side of the one, equal to two angles and the included side of the other; they are, therefore, equal (Prop. VII.); and the angle BFG is equal to the angle AEG. But AEG is, by construction, a right angle, whence BFG is also a right angle; that is, the two straight lines EC, FD are perpendicular to the same straight line, and are consequently parallel (Prop. XX.). Therefore, if a straight line, &c.

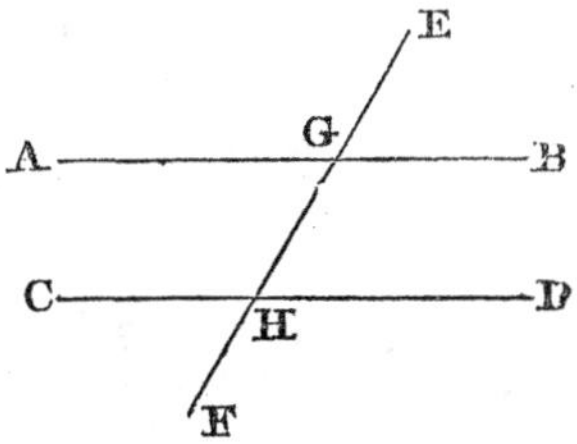

Scholium. When a straight line intersects two parallel lines, the *interior angles on the same side*, are those which lie within the parallels, and on the same side of the secant line, as AGH, GHC; also, BGH, GHD.

Alternate angles lie within the parallels, on different sides of the secant line, and are not adjacent to each other, as AGH GHD; also, BGH, GHC.

Either angle without the parallels being called an *exterior* angle, the *interior and opposite angle* on the same side, lies within the parallels, on the same side of the secant line, but not adjacent; thus, GHD is an interior angle opposite to the exterior angle EGB; so, also, with the angles CHG, AGE.

PROPOSITION XXII. THEOREM.

If a straight line, intersecting two other straight lines, makes the alternate angles equal to each other, or makes an exterior angle equal to the interior and opposite upon the same side of the secant line, these two lines are parallel.

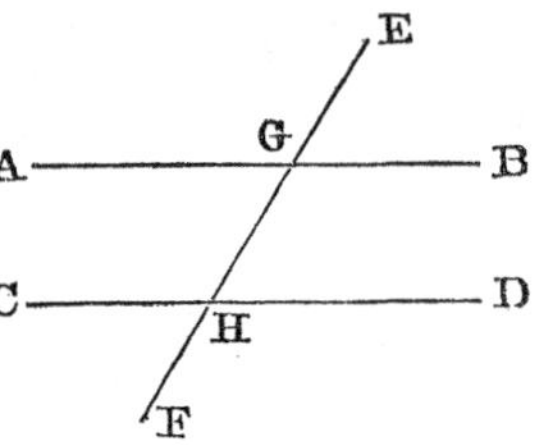

Let the straight line EF, which intersects the two straight lines AB, CD, make the alternate angles AGH, GHD equal to each other; then AB is parallel to CD. For, to each of the equal angles AGH, GHD, add the angle HGB; then the sum of AGH and HGB will be equal to the sum of GHD and HGB. But AGH and HGB are equal to two right angles (Prop. II.); therefore, GHD and HGB are equal to two right angles; and hence AB is parallel to CD (Prop. XXI.).

Again, if the exterior angle EGB is equal to the interior and opposite angle GHD, then is AB parallel to CD. For, the angle AGH is equal to the angle EGB (Prop. V.); and, by supposition, EGB is equal to GHD; therefore the angle AGH is equal to the angle GHD, and they are alternate angles; hence, by the first part of the proposition, AB is parallel to CD. Therefore, if a straight line, &c.

PROPOSITION XXIII. THEOREM.

(*Converse of Propositions XXI. and XXII.*)

If a straight line intersect two parallel lines, it makes the alternate angles equal to each other; also, any exterior angle equal to the interior and opposite on the same side; and the two interior angles on the same side together equal to two right angles.

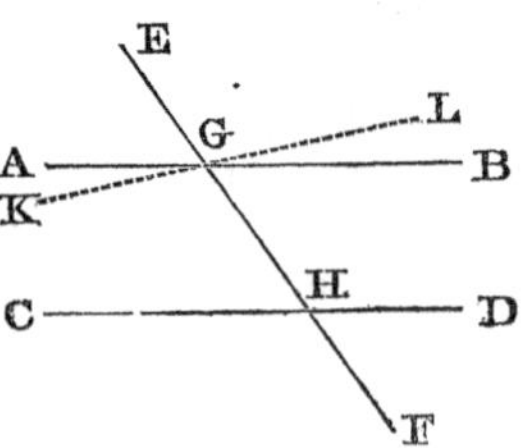

Let the straight line EF intersect the two parallel lines AB, CD; the alternate angles AGH, GHD are equal to each other; the exterior angle EGB is equal to the interior and opposite angle on the same side, GHD; and the two interior angles on the same side, BGH, GHD, are together equal to two right angles

For if AGH is not equal to GHD, through G draw the line KL, making the angle KGH equal to GHD; then KL must be parallel to CD (Prop. XXII.). But, by supposition, AB is parallel to CD; therefore, through the same point, G, two straight lines have been drawn parallel to CD, which is impossible (Axiom 12). Therefore, the angles AGH, GHD are not unequal, that is, they are equal to each other. Now the angle AGH is equal to EGB (Prop. V.), and AGH has been proved equal to GHD; therefore, EGB is also equa to GHD. Add to each of these equals the angle BGH; then will the sum of EGB, BGH be equal to the sum of BGH, GHD. But EGB, BGH are equal to two right angles (Prop. II.); therefore, also, BGH, GHD are equal to two right an gles. Therefore, if a straight line, &c

Cor. 1. If a straight line is perpendicular to one of two parallel lines, it is also perpendicular to the other.

Cor. 2. If two lines, KL and CD, make with EF the two angles KGH, GHC together less than two right angles, then will KL and CD meet, if sufficiently produced.

For if they do not meet, they are parallel (Def. 12). But they are not parallel; for then the angles KGH, GHC would be equal to two right angles.

PROPOSITION XXIV. THEOREM.

Straight lines which are parallel to the same line, are parallel to each other.

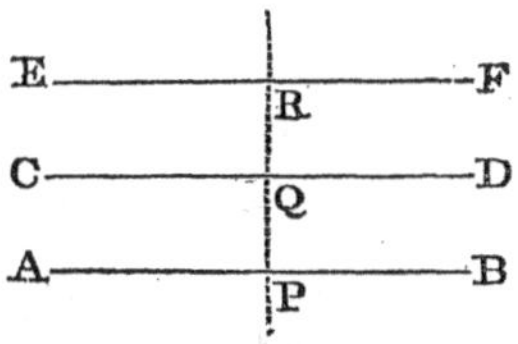

Let the straight lines AB, CD be each of them parallel to the line EF; then will AB be parallel to CD.

For, draw any straight line, as PQR, perpendicular to EF. Then, since AB is parallel to EF, PR, which is perpendicular to EF, will also be perpendicular to AB (Prop. XXIII., Cor. 1); and since CD is parallel to EF, PR will also be perpendicular to CD. Hence, AB and CD are both perpendicular to the same straight line, and are consequently parallel (Prop. XX.). Therefore, straight lines which are parallel, &c.

PROPOSITION XXV. THEOREM.

Two parallel straight lines are every where equally distant from each other.

Let AB, CD be two parallel straight lines. From any

points, E and F, in one of them, draw the lines EG, FH perpendicular to AB; they will also be perpendicular to CD (Prop. XXIII., Cor. 1). Join EH; then, because EG and FH are perpendicular to the same straight line AB they are parallel (Prop. XX.); therefore, the alternate angles, EHF, HEG, which they make with HE are equal (Prop. XXIII.). Again, because AB is parallel to CD, the alternate angles GHE, HEF are also equal. Therefore, the triangles HEF, EHG have two angles of the one equal to two angles of the other, each to each, and the side EH included between the equal angles, common; hence the triangles are equal (Prop. VII.); and the line EG, which measures the distance of the parallels at the point E, is equal to the line FH, which measures the distance of the same parallels at the point F. Therefore, two parallel straight lines, &c.

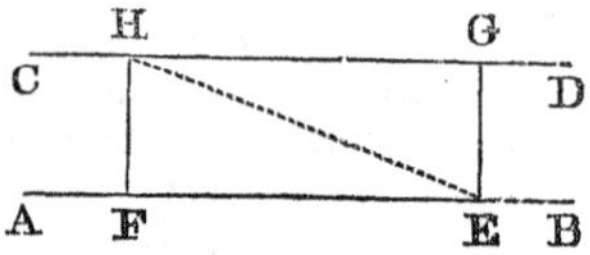

PROPOSITION XXVI. THEOREM.

Two angles are equal, when their sides are parallel, each to each, and are similarly situated.

Let BAC, DEF be two angles, having the side BA parallel to DE, and AC to EF; the two angles are equal to each other.

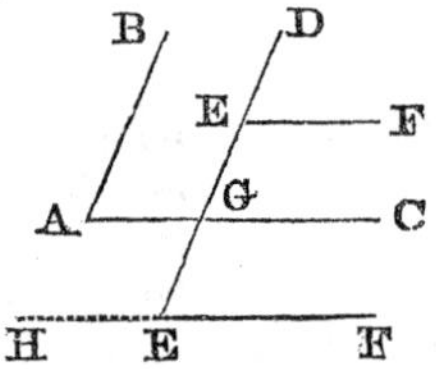

Produce DE, if necessary, until it meets AC in G. Then, because EF is parallel to GC, the angle DEF is equal to DGC (Prop. XXIII.); and because DG is parallel to AB, the angle DGC is equal to BAC; hence the angle DEF is equal to the angle BAC (Axiom 1). Therefore, two angles, &c.

Scholium. This proposition is restricted to the case in which the sides which contain the angles are similarly situated; because, if we produce FE to H, the angle DEH has its sides parallel to those of the angle BAC; but the two angles are not equal.

PROPOSITION XXVII. THEOREM.

If one side of a triangle is produced, the exterior angle is equal to the sum of the two interior and opposite angles; and the three interior angles of every triangle are equal to two right angles.

Let ABC be any plane triangle, and let the side BC be

produced to D; then will the exterior angle ACD be equal to the sum of the two interior and opposite angles A and B; and the sum of the three angles ABC, BCA, CAB is equal to two right angles.

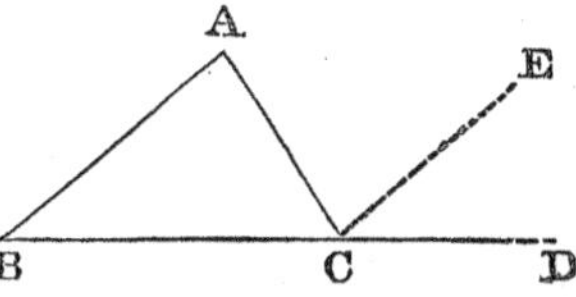

For, conceive CE to be drawn parallel to the side AB of the triangle; then, because AB is parallel to CE, and AC meets them, the alternate angles BAC, ACE are equal (Prop. XXIII.). Again, because AB is parallel to CE, and BD meets them, the exterior angle ECD is equal to the interior and opposite angle ABC. But the angle ACE was proved equal to BAC; therefore the whole exterior angle ACD is equal to the two interior and opposite angles CAB, ABC (Axiom 2). To each of these equals add the angle ACB; then will the sum of the two angles ACD, ACB be equal to the sum of the three angles ABC, BCA, CAB. But the angles ACD, ACB are equal to two right angles (Prop. II.); hence, also, the angles ABC, BCA, CAB are together equal to two right angles. Therefore, if one side of a triangle, &c.

Cor. 1. If the sum of two angles of a triangle is given, the third may be found by subtracting this sum from two right angles.

Cor. 2. If two angles of one triangle are equal to two angles of another triangle, the third angles are equal, and the triangles are mutually equiangular.

Cor. 3. A triangle can have but one right angle; for if there were two, the third angle would be nothing. Still less can a triangle have more than one obtuse angle.

Cor. 4. In a right-angled triangle, the sum of the two acute angles is equal to one right angle.

Cor. 5. In an equilateral triangle, each of the angles is one third of two right angles, or two thirds of one right angle.

PROPOSITION XXVIII. THEOREM.

The sum of all the interior angles of a polygon, is equal to twice as many right angles, wanting four, as the figure has sides.

Let ABCDE be any polygon; then the sum of all its interior angles A, B, C, D, E is equal to twice as many right angles, wanting four, as the figure has sides (see next page).

For, from any point, F, within it, draw lines FA, FB, FC, &c., to all the angles. The polygon is thus divided into as many triangles as it has sides. Now the sum of the three angles of each of these triangles, is equal to two right angles (Prop. XXVII.); therefore the sum of the angles of all the triangles is equal to twice as many right angles as the polygon has sides. But the same angles are equal to the angles of the polygon, together with the angles at the point F, that is, together with four right angles (Prop. V., Cor. 2). Therefore the angles of the polygon are equal to twice as many right angles as the figure has sides, wanting four right angles.

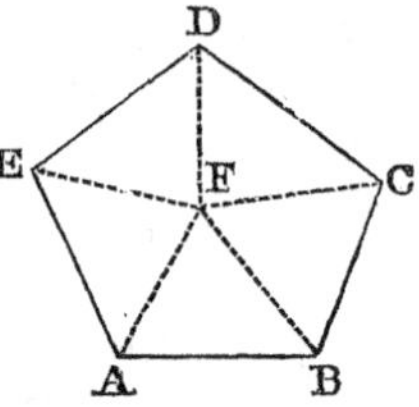

Cor. 1. The sum of the angles of a quadrilateral is four right angles; of a pentagon, six right angles; of a hexagon, eight, &c.

Cor. 2. *All the exterior angles of a polygon are together equal to four right angles.* Because every interior angle, ABC, together with its adjacent exterior angle, ABD, is equal to two right angles (Prop. II.); therefore the sum of all the interior and exterior angles, is equal to twice as many right angles as the polygon has sides; that is, they are equal to all the interior angles of the polygon, together with four right angles. Hence the sum of the exterior angles must be equal to four right angles (Axiom 3).

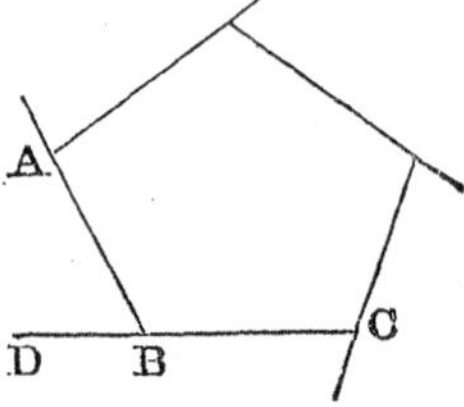

PROPOSITION XXIX. THEOREM.

The opposite sides and angles of a parallelogram are equal to each other.

Let ABDC be a parallelogram; then will ts opposite sides and angles be equal to each other.

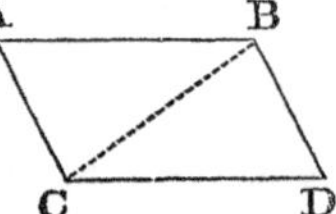

Draw the diagonal BC; then, because AB s parallel to CD, and BC meets them, the alternate angles ABC, BCD are equal to each other (Prop. XXIII.). Also, because AC is parallel to BD, and BC meets them, the alternate angles BCA, CBD are equal to each other. Hence the two triangles ABC, BCD have two angles, ABC, BCA of the one, equal to two angles, BCD, CBD, of the other, each to each, and the side BC included between these equal angles, common to the two triangles; therefore their other sides are equal, each to each, and the third angle of the one to the third angle of the other (Prop. VII.), viz.

the side AB to the side CD, and AC to BD, and the angle BAC equal to the angle BDC. Also, because the angle ABC is equal to the angle BCD, and the angle CBD to the angle BCA, the whole angle ABD is equal to the whole angle ACD. But the angle BAC has been proved equal to the angle BDC; therefore the opposite sides and angles of a parallelogram are equal to each other.

Cor. Two parallels, AB, CD, comprehended between two other parallels, AC, BD, are equal; and the diagonal BC divides the parallelogram into two equal triangles.

PROPOSITION XXX. THEOREM (*Converse of Prop. XXIX.*).

If the opposite sides of a quadrilateral are equal, each to each, the equal sides are parallel, and the figure is a parallelogram.

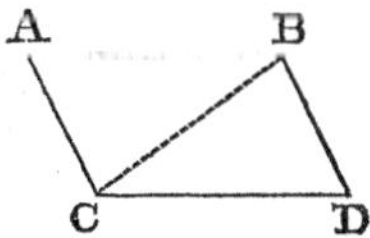

Let ABDC be a quadrilateral, having its opposite sides equal to each other, viz.: the side AB equal to CD, and AC to BD; then will the equal sides be parallel, and the figure will be a parallelogram.

Draw the diagonal BC; then the triangles ABC, BCD have all the sides of the one equal to the corresponding sides of the other, each to each; therefore the angle ABC is equal to the angle BCD (Prop. XV.), and, consequently, the side AB is parallel to CD (Prop. XXII.). For a like reason, AC is parallel to BD; hence the quadrilateral ABDC is a parallelogram. Therefore, if the opposite sides, &c.

PROPOSITION XXXI. THEOREM.

If two opposite sides of a quadrilateral are equal and parallel, the other two sides are equal and parallel, and the figure is a parallelogram.

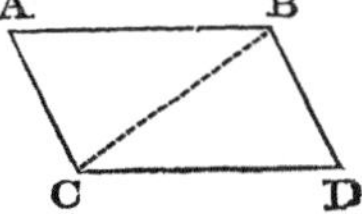

Let ABDC be a quadrilateral, having the sides AB, CD equal and parallel; then will the sides AC, BD be also equal and parallel, and the figure will be a parallelogram.

Draw the diagonal BC; then, because AB is parallel to CD, and BC meets them, the alternate angles ABC, BCD are equal (Prop. XXIII). Also, because AB is equal to CD, and BC is common to the two triangles ABC BCD, the two triangles ABC, BCD have two sides and the included angle of the one, equal to two sides and the included angle of the other; therefore, the side AC is equal to BD (Prop. VI.), and the angle ACB to the angle CBD And, because the straight line BC meets the two straight lines AC, BD, making the alternate angles BCA, CBD equal to each other, AC is parallel to BD (Prop. XXII.); hence the figure ABDC is a parallelogram. Therefore, if two opposite sides, &c.

PROPOSITION XXXII. THEOREM.

The diagonals of every parallelogram bisect each other

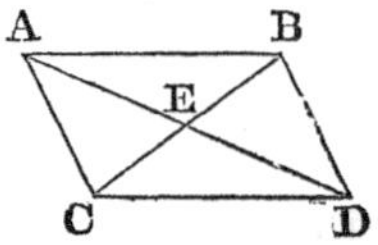

Let ABDC be a parallelogram whose diagonals, AD, BC, intersect each other in E; then will AE be equal to ED, and BE to EC.

A B E C D

Because the alternate angles ABE, ECD are equal (Prop. XXIII.), and also the alternate angles EAB, EDC, the triangles ABE, DCE have two angles in the one equal to two angles in the other, each to each, and the included sides AB, CD are also equal; hence the remaining sides are equal, viz.: AE to ED, and CE to EB. Therefore, the diagonals of every parallelogram, &c.

Cor. If the side AB is equal to AC, the triangles AEB, AEC have all the sides of the one equal to the corresponding sides of the other, and are consequently equal; hence the angle AEB will equal the angle AEC, and therefore *the diagonals of a rhombus bisect each other at right angles*

BOOK II.

RATIO AND PROPORTION.

On the Relation of Magnitudes to Numbers.

THE ratios of magnitudes may be expressed by numbers either exactly or approximately; and in the latter case, the approximation can be carried to any required degree of precision.

Thus, let it be proposed to find the numerical ratio of two straight lines, AB and CD.

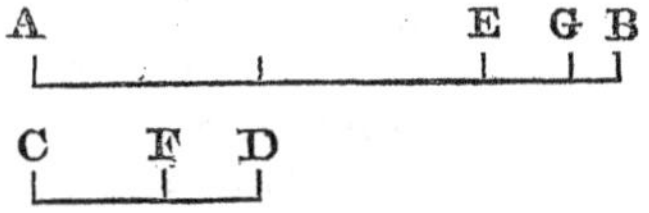

From the greater line AB, cut off a part equal to the less, CD, as many times as possible; for example, twice, with a remainder EB. From CD, cut off a part equal to the remainder EB as often as possible; for example, once, with a remainder FD. From the first remainder, BE, cut off a part equal to FD as often as possible; for example, once, with a remainder GB. From the second remainder, FD, cut off a part equal to the third, GB, as many times as possible. Continue this process until a remainder is found which is contained an exact number of times in the preceding one. This last remainder will be the common measure of the proposed lines; and regarding it as the measuring unit, we may easily find the values of the preceding remainders, and at length those of the proposed lines; whence we obtain their ratio in numbers.

A E G B

C F D

For example, if we find GB is contained exactly twice in FD, GB will be the common measure of the two proposed lines. Let GB be called unity, then FD will be equal to 2. But EB contains FD once, plus GB; therefore, EB=3. CD contains EB once, plus FD; therefore, CD=5. AB contains CD twice, plus EB; therefore, AB=13. Consequently, the ratio of the two lines AB, CD is that of 13 to 5.

However far the operation is continued, it is possible that we may never find a remainder which is contained an exact number of times in the preceding one. In such cases, the *ex-*

act ratio can not be expressed in numbers; but, by taking the measuring unit sufficiently small, a ratio may always be found, which shall approach as near as we please to the true ratio.

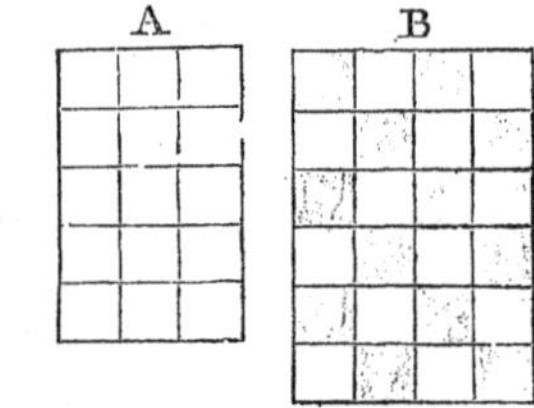

So, also, in comparing two surfaces, we seek some unit of measure which is contained an exact number of times in each of them. Let A and B represent two surfaces, and let a square inch be the unit of measure. Now, if this measuring unit is contained 15 times in A and 24 times in B, then the ratio of A to B is that of 15 to 24. And although it may be difficult to find this measuring unit, we may still conceive it to exist; or, if there is no unit which is contained an exact number of times in both surfaces, yet, since the unit may be made as small as we please, we may represent their ratio in numbers to any degree of accuracy required.

Again, if we wish to find the ratio of two solids, A and B, we seek some unit of measure which is contained an exact number of times in each of them. If we take a cubic inch as the unit of measure, and we find it to be contained 9 times in A, and 13 times in B, then the ratio of A to B is the same as that of 9 to 13. And even if there is no unit which is contained an exact number of times in both solids, still, by taking the unit sufficiently small, we may represent their ratio in numbers to any required degree of precision.

Hence the ratio of two magnitudes in geometry, is the same as the ratio of two numbers, and thus each magnitude has its *numerical representative.* We therefore conclude that ratio in geometry is essentially the same as in arithmetic, and we might refer to our treatise on algebra for such properties of ratios as we have occasion to employ. However, in order to render the present treatise complete in itself, we will here demonstrate the most useful properties.

Definitions.

Def. 1. ***Ratio*** is the relation which one magnitude bears to another with respect to quantity.

Thus, the ratio of a line two inches in length, to another six inches in length is denoted by 2 divided by 6, *i. e.*, $\frac{2}{6}$ or $\frac{1}{3}$, the number 2 being the third part of 6. So, also, the ratio of 3 feet to 6 feet is expressed by $\frac{3}{6}$ or $\frac{1}{2}$.

A ratio is most conveniently written as a fraction; thus,

the ratio of A to B is written $\frac{A}{B}$. The two magnitudes compared together are called the *terms* of the ratio; the first is called the *antecedent*, and the second the *consequent*.

Def. 2. *Proportion is an equality of ratios.*

Thus, if A has to B the same ratio that C has to D, these four quantities form a proportion, and we write it

$$\frac{A}{B}=\frac{C}{D},$$

or

$$A : B : C : D.$$

The first and last terms of a proportion are called the two *extremes*, and the second and third terms the two *means*.

Of four proportional quantities, the last is called a *fourth proportional* to the other three, taken in order.

Since

$$\frac{A}{B}=\frac{C}{D},$$

it is obvious that if A is greater than B, C must be greater than D; if equal, equal; and if less, less; that is, if one antecedent is greater than its consequent, the other antecedent must be greater than its consequent; if equal, equal; and if less, less.

Def. 3. Three quantities are said to be proportional, when the ratio of the first to the second is equal to the ratio of the second to the third; thus, if A, B, and C are in proportion, then

$$A : B : : B : C.$$

In this case the middle term is said to be a *mean proportional* between the other two.

Def. 4. Two magnitudes are said to be *equimultiples* of two others, when they contain those others the same number of times exactly. Thus, 7A, 7B are equimultiples of A and B; so, also, are mA and mB.

Def. 5. The ratio of B to A is said to be the *reciprocal* of the ratio of A to B.

Def. 6. *Inversion* is when the antecedent is made the consequent, and the consequent the antecedent.

Thus, if $A : B : \cdot C : D$;

then, inversely,

$$B : A . : D : C.$$

Def. 7. *Alternation* is when antecedent is compared with antecedent, and consequent with consequent.

Thus, if $A : B : : C : D$;

then, by alternation,

$$A : C : : B : D.$$

Def. 8. *Composition* is when the sum of antecedent and consequent is compared either with the antecedent or consequent.

Thus, if $A : B :: C : D$;
then, by composition,

$$A+B : A :: C+D : C, \text{ and } A+B : B :: C+D : D.$$

Def. 9. *Division* is when the difference of antecedent and consequent is compared either with the antecedent or consequent.

Thus, if $A : B :: C : D$;
then, by division,

$$A-B : A :: C-D : C, \text{ and } A-B : B :: C-D : D.$$

Axioms.

1. Equimultiples of the same, or equal magnitudes, are equal to each other.
2. Those magnitudes of which the same or equal magnitudes are equimultiples, are equal to each other.

PROPOSITION I. THEOREM.

If four quantities are proportional, the product of the two extremes is equal to the product of the two means.

It has been shown that the ratio of two magnitudes, whether they are lines, surfaces, or solids, is the same as that of two numbers, which we call their *numerical representatives.*

Let, then, A, B, C, D be the numerical representatives of four proportional quantities, so that $A : B :: C : D$; then will $A \times D = B \times C$.

For, since the four quantities are proportional,

$$\frac{A}{B} = \frac{C}{D}.$$

Multiplying each of these equal quantities by B (Axiom 1), we obtain

$$A = \frac{B \times C}{D}.$$

Multiplying each of these last equals by D, we have

$$A \times D = B \times C.$$

Cor. If there are three proportional quantities, the product of the two extremes is equal to the square of the mean.

Thus, if $A : B :: B : C$;
then, by the proposition,

$$A \times C = B \times B, \text{ which is equal to } B^2.$$

PROPOSITION II. THEOREM (*Converse of Prop. I.*).

If the product of two quantities is equal to the product of two other quantities, the first two may be made the extremes, and the other two the means of a proportion.

Thus, suppose we have $A \times D = B \times C$; then will

$$A : B :: C : D.$$

For, since $A \times D = B \times C$, dividing each of these equals by D (Axiom 2), we have

$$A = \frac{B \times C}{D}.$$

Dividing each of these last equals by B, we obtain

$$\frac{A}{B} = \frac{C}{D},$$

that is, the ratio of A to B is equal to that of C to D, or,

$$A : B :: C : D.$$

PROPOSITION III. THEOREM.

If four quantities are proportional, they are also proportional when taken alternately.

Let A, B, C, D be the numerical representatives of four proportional quantities, so that $A : B :: C : D$; then will

$$A : C :: B : D.$$

For, since $A : B :: C : D$,
by Prop. I., $A \times D = B \times C$.
And, since $A \times D = B \times C$,
by Prop. II., $A : C :: B : D$.

PROPOSITION IV. THEOREM.

Ratios that are equal to the same ratio, are equal to each other.

Let $A : B :: C : D$,
and $A : B :: E : F$;
then will $C : D :: E : F$.
For, since $A : B :: C : D$,
we have $\frac{A}{B} = \frac{C}{D}$.

And, since $A : B :: E \quad F$,

we have $\frac{A}{B} = \frac{E}{F}$.

But $\frac{C}{D}$ and $\frac{E}{F}$, being severally equal to $\frac{A}{B}$, must be equal to each other, and therefore

$$C : D :: E : F.$$

Cor. If the antecedents of one proportion are equal to the antecedents of another proportion, the consequents are proportional.

If $A : B :: C : D$,
and $A : E :: C : F$;
then will $B : D :: E : F$.

For, by alternation (Prop. III.), the first proportion becomes

$$A : C :: B : D,$$

and the second, $A : C :: E : F$.

Therefore, by the proposition,

$$B : D :: E : F.$$

PROPOSITION V. THEOREM.

If four quantities are proportional, they are also proportional when taken inversely.

Let $A : B :: C : D$;
then will $B : A :: D : C$.
For, since $A : B :: C : D$,
by Prop. I., $A \times D = B \times C$,
or, $B \times C = A \times D$;
therefore, by Prop. II.,

$$B : A :: D : C.$$

PROPOSITION VI. THEOREM.

If four quantities are proportional, they are also proportional by composition.

Let $A : B :: C : D$,
then will $A+B : A :: C+D . C$.
For, since $A : B :: C : D$,
by Prop. I., $B \times C = A \times D$.
To each of these equals add

$$A \times C = A \times C,$$

then $A \times C + B \times C = A \times C + A \times D$,

$$(A+B)\times C=A\times(C+D).$$

Therefore, by Prop. II.,

$$A+B : A :: C+D : C.$$

PROPOSITION VII. THEOREM.

If four quantities are proportional, they are also proportional by division.

Let $A : B :: C : D$;

then will $A-B : A :: C-D : C.$

For, since $A : B :: C : D,$

by Prop. I., $B\times C=A\times D.$

Subtract each of these equals from $A\times C$;

then $A\times C-B\times C=A\times C-A\times D,$

or, $(A-B)\times C=A\times(C-D).$

Therefore, by Prop. II.,

$$A-B : A :: C-D : C.$$

Cor. $A+B : A-B :: C+D : C-D.$

PROPOSITION VIII. THEOREM.

Equimultiples of two quantities have the same ratio as the quantities themselves.

Let A and B be any two quantities, and mA, mB their equimultiples; then will

$$A : B :: mA : mB.$$

For $m\times A\times B=m\times A\times B,$

or, $A\times mB=B\times mA.$

Therefore, by Prop. II.,

$$A : B :: mA : mB.$$

PROPOSITION IX. THEOREM.

If any number of quantities are proportional, any one antecedent is to its consequent, as the sum of all the antecedents, is to the sum of all the consequents.

Let $A : B :: C : D :: E : F$, &c.;

then will $A : B :: A+C+E : B+D+F$

For, since $A : B :: C : D,$

we have $A\times D=B\times C.$

And, since $A : B :: E : F,$

we have $A\times F=B\times E.$

To these equals add

$$A\times B=A\times B.$$

and we have

$$A\times B+A\times D+A\times F=A\times B+B\times C+B\times E;$$

or, $$A\times(B+D+F)=B\times(A+C+E).$$

Therefore, by Prop. II.,

$$A : B :: A+C+E : B+D+F.$$

PROPOSITION X. THEOREM.

If four quantities are proportional, their squares or cubes are also proportional.

Let $A : B :: C : D;$

then will $A^2 : B^2 :: C^2 : D^2,$

and $A^3 : B^3 :: C^3 : D^3.$

For, since $A : B :: C : D,$

by Prop. I., $A\times D=B\times C;$

or, multiplying each of these equals by itself (Axiom 1), we have

$$A^2\times D^2=B^2\times C^2;$$

and multiplying these last equals by $A\times D=B\times C$, we have

$$A^3\times D^3=B^3\times C^3.$$

Therefore, by Prop. II.,

$$A^2 : B^2 :: C^2 : D^2,$$

and $A^3 : B^3 :: C^3 : D^3.$

PROPOSITION XI. THEOREM.

If there are two sets of proportional quantities, the products of the corresponding terms are proportional.

Let $A : B :: C : D,$

and $E : F :: G : H;$

then will $A\times E : B\times F :: C\times G : D\times H.$

For, since $A : B :: C : D,$

by Prop. I., $A\times D=B\times C.$

And, since $E : F :: G : H,$

by Prop. I., $E\times H=F\times G.$

Multiplying together these equal quantities, we have

$$A\times D\times E\times H=B\times C\times F\times G;$$

or, $$(A\times E)\times(D\times H)=(B\times F)\times(C\times G);$$

therefore, by Prop. II.,

$$A\times E : B\times F :: C\times G : D\times H.$$

Cor. If $A : B :: C : D,$

and $B : F :: G : H;$

then $A : F :: C\times G : D\times H.$

For, by the proposition,

$$A\times B : B\times F :: C\times G : D\times H.$$

Also, by Prop. VIII.,

$$A\times B : B\times F :: A : F;$$

hence, by Prop. IV.,

$$A : F :: C\times G : D\times H.$$

PROPOSITION XII. THEOREM.

If three quantities are proportional, the first is to the third, as the square of the first to the square of the second.

Thus, if $A : B :: B : C$;
then $A : C :: A^2 : B^2$.
For, since $A : B :: B : C$,
and $A : B :: A : B$;
therefore, by Prop. XI.,

$$A^2 : B^2 :: A\times B : B\times C.$$

But, by Prop. VIII.,

$$A\times B : B\times C :: A : C;$$

hence, by Prop. IV., $A : C :: A^2 : B^2$.

BOOK III.

THE CIRCLE, AND THE MEASURE OF ANGLES.

Definitions.

1. A *circle* is a plane figure bounded by a line, every point of which is equally distant from a point within, called the *center.*

This bounding line is called the *circumference* of the circle.

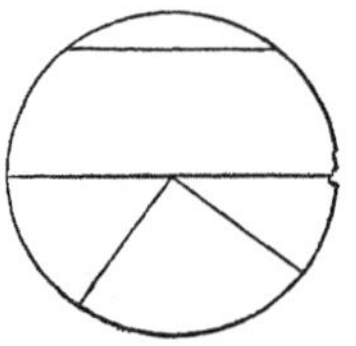

2. A *radius* of a circle is a straight line drawn from the center to the circumference.

A *diameter* of a circle is a straight line passing through the center, and terminated both ways by the circumference.

Cor. All the radii of a circle are equal; all the diameters are equal also, and each double of the radius.

3. An *arc* of a circle is any part of the circumference.

The *chord* of an arc is the straight line which joins its two extremities.

4. A *segment* of a circle is the figure included between an arc and its chord.

5. A *sector* of a circle is the figure included between an arc, and the two radii drawn to the extremities of the arc.

6. A straight line is said to be *inscribed* in a circle, when its extremities are on the circumference.

An *inscribed angle* is one whose sides are inscribed.

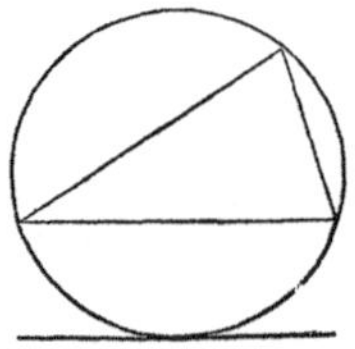

7. A polygon is said to be *inscribed* in a circle, when all its sides are inscribed. The circle is then said to be *described* about the polygon.

8. A *secant* is a line which cuts the circumference, and lies partly within and partly without the circle.

9. A straight line is said to *touch* a circle, when it meets the circumference, and, being produced, does not cut it. Such a line is called a *tangent*, and the point in which it meets the circumference, is called the *point of contact.*

10. Two circumferences *touch* each other when they meet, but do not cut one another.

11. A polygon is *described* about a circle, when each side of the polygon touches the circumference of the circle.

In the same case, the circle is said to be *inscribed* in the polygon.

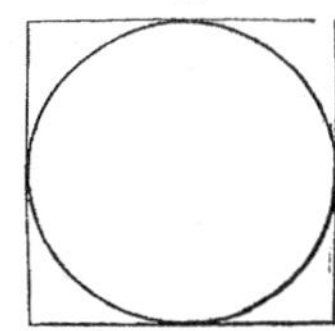

PROPOSITION I. THEOREM.

Every diameter divides the circle and its circumference into two equal parts.

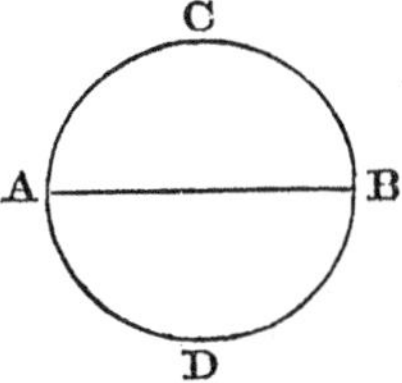

Let ACBD be a circle, and AB its diameter. The line AB divides the circle and its circumference into two equal parts. For, if the figure ADB be applied to the figure ACB, while the line AB remains common to both, the curve line ACB must coincide exactly with the curve line ADB. For, if any part of the curve ACB were to fall either within or without the curve ADB, there would be points in one or the other unequally distant from the center which is contrary to the definition of a circle. Therefore, every diameter, &c.

PROPOSITION II. THEOREM.

A straight line can not meet the circumference of a circle in more than two points.

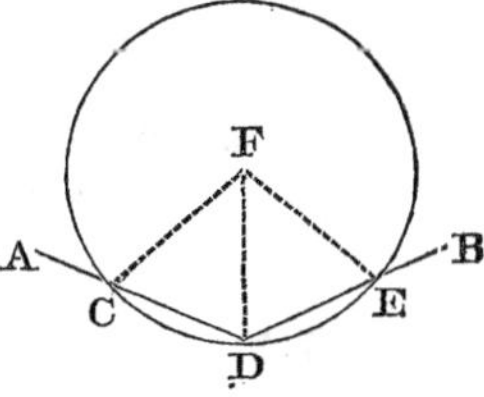

For, if it is possible, let the straight line ADB meet the circumference CDE in three points, C, D, E. Take F, the center of the circle, and join FC, FD, FE. Then, because F is the center of the circle, the three straight lines FC, FD, FE are all equal to each other; hence, three equal straight lines have been drawn from the same point to the same straight line.

which is impossible (Prop. XVII., Cor. 2, Book I.). Therefore, a straight line, &c.

PROPOSITION III. THEOREM.

In equal circles, equal arcs are subtended by equal chords and, conversely, equal chords subtend equal arcs.

Let ADB, EHF be equal circles, and let the arcs AID, EMH also be equal; then will the chord AD be equal to the chord EH.

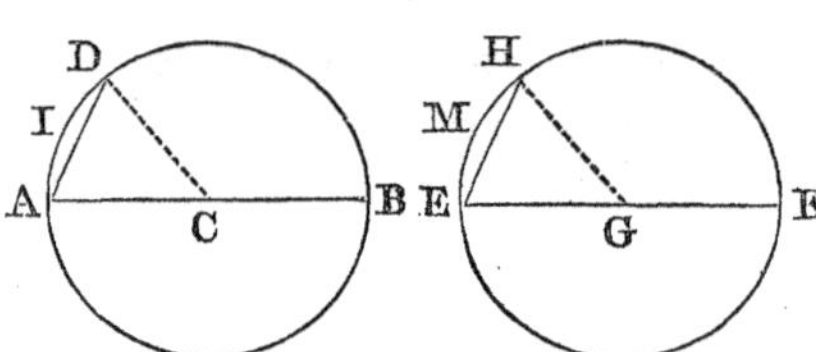

For, the diameter AB being equal to the diameter EF, the semicircle ADB may be applied exactly to the semicircle EHF, and the curve line AIDB will coincide entirely with the curve line EMHF (Prop. I.). But the arc AID is, by hypothesis, equal to the arc EMH; hence the point D will fall on the point H, and therefore the chord AD is equal to the chord EH (Axiom 11, B. I.).

Conversely, if the chord AD is equal to the chord EH, then the arc AID will be equal to the arc EMH.

For, if the radii CD, GH are drawn, the two triangles ACD, EGH will have their three sides equal, each to each viz.: AC to EG, CD to GH, and AD equal to EH; the triangles are consequently equal (Prop. XV., B. I.), and the angle ACD is equal to the angle EGH. Let, now, the semicircle ADB be applied to the semicircle EHF, so that AC may coincide with EG; then, since the angle ACD is equal to the angle EGH, the radius CD will coincide with the radius GH, and the point D with the point H. Therefore, the arc AID must coincide with the arc EMH, and be equal to it. Hence, in equal circles, &c.

PROPOSITION IV. THEOREM.

In equal circles, equal angles at the center, are subtended by equal arcs; and, conversely, equal arcs subtend equal angles at the center.

Let AGB, DHE be two equal circles, and let ACB, DFE be equal angles at their centers; then will the arc AB be equal to the arc DE. Join AB, DE; and, because the cir

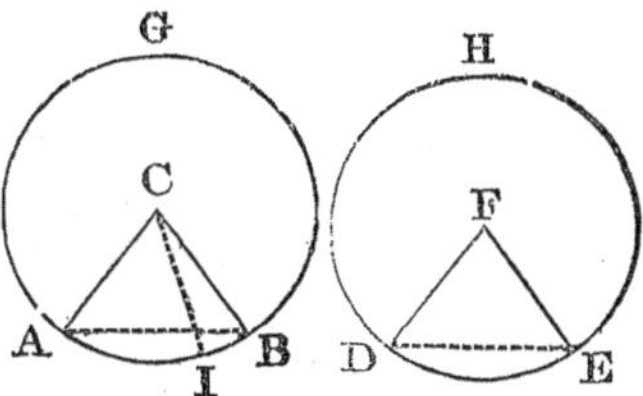

cles AGB, DHE are equal, their radii are equal. Therefore, the two sides CA, CB are equal to the two sides FD, FE; also, the angle at C is equal to the angle at F; therefore, the base AB is equal to the base DE (Prop. VI., B. I.). And, because the chord AB is equal to the chord DE, the arc AB must be equal to the arc DE (Prop. III.).

Conversely, if the arc AB is equal to the arc DE, the angle ACB will be equal to the angle DFE. For, if these angles are not equal, one of them is the greater. Let ACB be the greater, and take ACI equal to DFE; then, because equal angles at the center are subtended by equal arcs, the arc AI is equal to the arc DE. But the arc AB is equal to the arc DE; therefore, the arc AI is equal to the arc AB, the less to the greater, which is impossible. Hence the angle ACB is not unequal to the angle DFE, that is, it is equal to it. Therefore, in equal circles, &c.

PROPOSITION V. THEOREM.

In the same circle, or in equal circles, a greater arc is subtended by a greater chord; and, conversely, the greater chord subtends the greater arc.

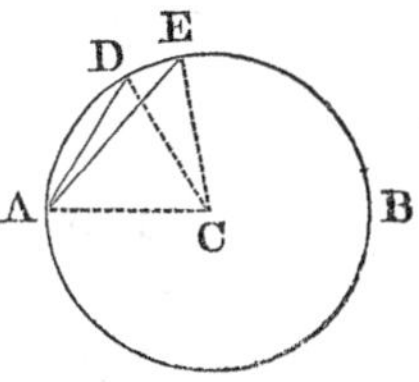

In the circle AEB, let the arc AE be greater than the arc AD; then will the chord AE be greater than the chord AD.

Draw the radii CA, CD, CE. Now, if the arc AE were equal to the arc AD, the angle ACE would be equal to the angle ACD (Prop. IV.); hence it is clear that if the arc AE be greater than the arc AD, the angle ACE must be greater than the angle ACD. But the two sides AC, CE of the triangle ACE are equal to the two AC, CD of the triangle ACD, and the angle ACE is greater than the angle ACD; therefore, the third side AE is greater than the third side AD (Prop. XIII., B. I.); hence the chord which subtends the greater arc is the greater.

Conversely, if the chord AE is greater than the chord AD, the arc AE is greater than the arc AD. For, because the two triangles ACE, ACD have two sides of the one equal to two sides of the other, each to each, but the base AE of the one is greater than the base AD of the other, therefore the angle ACE is greater than the angle ACD (Prop. XIV. B. I.); and hence the arc AE is greater than the arc AD (Prop. IV.). Therefore, in the same circle, &c.

Scholium. The arcs here treated of are supposed to be less than a semicircumference. If they were greater, the opposite property would hold true, that is, the greater the arc the smaller the chord.

PROPOSITION VI. THEOREM.

The radius which is perpendicular to a chord, bisects the chord, and also the arc which it subtends.

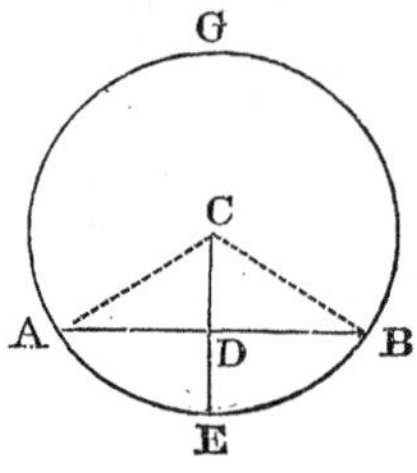

Let ABG be a circle, of which AB is a chord, and CE a radius perpendicular to it; the chord AB will be bisected in D, and the arc AEB will be bisected in E.

Draw the radii CA, CB. The two right-angled triangles CDA, CDB have the side AC equal to CB, and CD common; therefore the triangles are equal, and the base AD is equal to the base DB (Prop. XIX., B. I.).

Secondly, since ACB is an isosceles triangle, and the line CD bisects the base at right angles, it bisects also the vertical angle ACB (Prop. X., Cor. 1, B. I.). And, since the angle ACE is equal to the angle BCE, the arc AE must be equal to the arc BE (Prop. IV.); hence the radius CE, perpendicular to the chord AB, divides the arc subtended by this chord, into two equal parts in the point E. Therefore, the radius, &c.

Scholium. The center C, the middle point D of the chord AB, and the middle point E of the arc subtended by this chord, are three points situated in a straight line perpendicular to the chord. Now two points are sufficient to determine the position of a straight line; therefore any straight line which passes through two of these points, will necessarily pass through the third, and be perpendicular to the chord. Also, *the perpendicular at the middle of a chord passes through the center of the circle, and through the middle of the arc subtended by the chord.*

PROPOSITION VII. THEOREM.

Through three given points, not in the same straight line, one circumference may be made to pass, and but one.

Let A, B, C be three points not in the same straight line; they all lie in the circumference of the same circle. Join AB, AC, and bisect these lines by the perpendiculars DF, EF; DF and EF produced will meet one another. For, join DE; then, because the angles ADF, AEF are together equal to two right angles, the angles FDE and FED are together less than two right angles; therefore DF and EF will meet if produced (Prop. XXIII., Cor. 2, B. I.). Let them meet in F. Since this point lies in the perpendicular DF, it is equally distant from the two points A and B (Prop. XVIII., B. I.); and, since it lies in the perpendicular EF, it is equally distant from the two points A and C; therefore the three distances FA, FB, FC are all equal; hence the circumference described from the center F with the radius FA will pass through the three given points A, B, C.

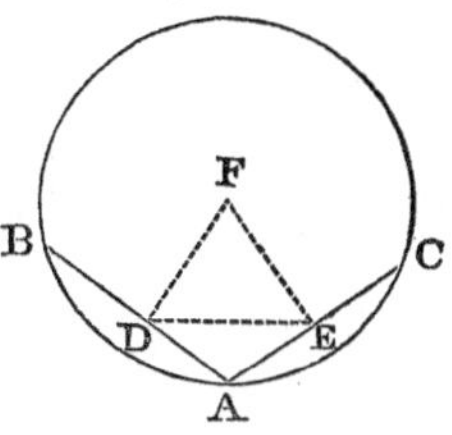

Secondly. No other circumference can pass through the same points. For, if there were a second, its center could not be out of the line DF, for then it would be unequally distant from A and B (Prop. XVIII., B. I.); neither could it be out of the line FE, for the same reason; therefore, it must be on both the lines DF, FE. But two straight lines can not cut each other in more than one point; hence only one circumference can pass through three given points. Therefore, through three given points, &c.

Cor. Two circumferences can not cut each other in more than two points, for, if they had three common points, they would have the same center, and would coincide with each other.

PROPOSITION VIII. THEOREM.

Equal chords are equally distant from the center; and of two unequal chords, the less is the more remote from the center.

Let the chords AB, DE, in the circle ABED, be equal to one another; they are equally distant from the center. Take

C, the center of the circle, and from it draw CF, CG, perpendiculars to AB, DE. Join CA, CD; then, because the radius CF is perpendicular to the chord AB, it bisects it (Prop. VI.). Hence AF is the half of AB; and, for the same reason, DG is the half of DE. But AB is equal to DE; therefore AF is equal to DG (Axiom 7, B. I.). Now, in the right-angled triangles ACF, DCG, the hypothenuse AC is equal to the hypothenuse DC, and the side AF is equal to the side DG; therefore the triangles are equal, and CF is equal to CG (Prop. XIX., B. I.); hence the two equal chords AB, DE are equally distant from the center.

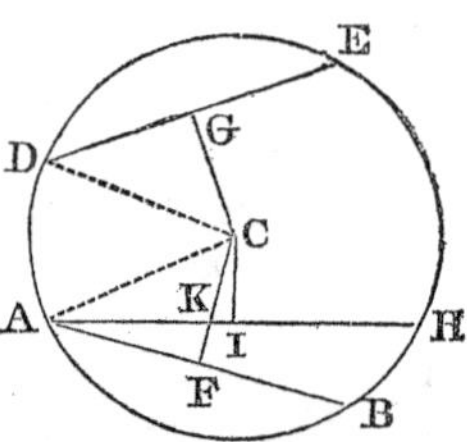

Secondly. Let the chord AH be greater than the chord DE; DE is further from the center than AH. For, because the chord AH is greater than the chord DE, the arc ABH is greater than the arc DE (Prop. V.). From the arc ABH cut off a part, AB, equal to DE; draw the chord AB, and let fall CF perpendicular to this chord, and CI perpendicular to AH. It is plain that CF is greater than CK, and CK than CI (Prop. XVII., B. I.); much more, then, is CF greater than CI. But CF is equal to CG, because the chords AB, DE are equal; hence CG is greater than CI. Therefore equal chords, &c.

Cor. Hence the diameter is the longest line that can be inscribed in a circle.

PROPOSITION IX. THEOREM.

A straight line perpendicular to a diameter at its extremity, is a tangent to the circumference.

Let ABG be a circle, the center of which is C, and the diameter AB; and let AD be drawn from A perpendicular to AB; AD will be a tangent to the circumference.

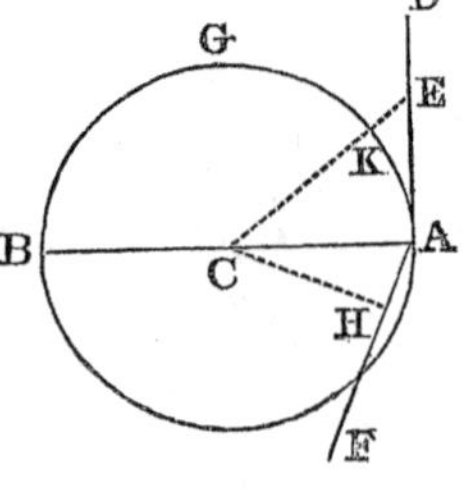

In AD take any point E, and join CE; then, since CE is an oblique line, it is longer than the perpendicular CA (Prop. XVII., B. I.). Now CA is equal to CK; therefore CE is greater than CK, and the point E must be without the circle. But E is any point whatever in the line AD; therefore AD has only the point A in common with the circumference, hence it is a tangent (Def. 9). Therefore, a straight line, &c.

Scholium. Through the same point A in the circumference, only one tangent can be drawn. For, if possible let a second tangent, AF, be drawn; then, since CA can not be perpendicular to AF (Prop. XVI., Cor., B. I.), another line, CH, must be perpendicular to AF, and therefore CH must be less than CA (Prop. XVII., B. I.; hence the point H falls within the circle, and AH produced will cut the circumference.

PROPOSITION X. THEOREM.

Two parallels intercept equal arcs on the circumference.

The proposition admits of three cases:

First. When the two parallels are secants, as AB, DE. Draw the radius CH perpendicular to AB; it will also be perpendicular to DE (Prop. XXIII., Cor. 1, B. I.); therefore, the point H will be at the same time the middle of the arc AHB, and of the arc DHE (Prop. VI.). Hence the arc DH is equal to the arc HE, and the arc AH equal to HB, and therefore the arc AD is equal to the arc BE (Axiom 3, B. I.).

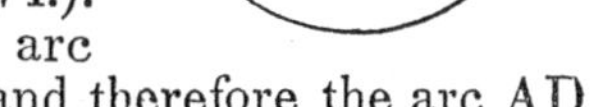

Second. When one of the two parallels is a secant, and the other a tangent. To the point of contact, H, draw the radius CH; it will be perpendicular to the tangent DE (Prop. IX.), and also to its parallel AB. But since CH is perpendicular to the chord AB, the point H is the middle of the arc AHB (Prop. VI.); therefore the arcs AH, HB, included between the parallels AB, DE, are equal.

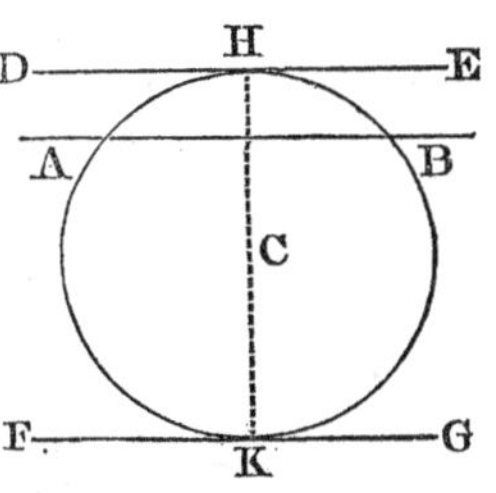

Third. If the two parallels DE, FG are tangents, the one at H, the other at K, draw the parallel secant AB; then, according to the former case, the arc AH is equal to HB, and the arc AK is equal to KB; hence the whole arc HAK is equal to the whole arc HBK (Axiom 2, B. I.). It is also evident that each of these arcs is a semicircumference. Therefore, two parallels, &c.

PROPOSITION XI. THEOREM.

If two circumferences cut each other, the chord which joins the points of intersection, is bisected at right angles by the straight line joining their centers.

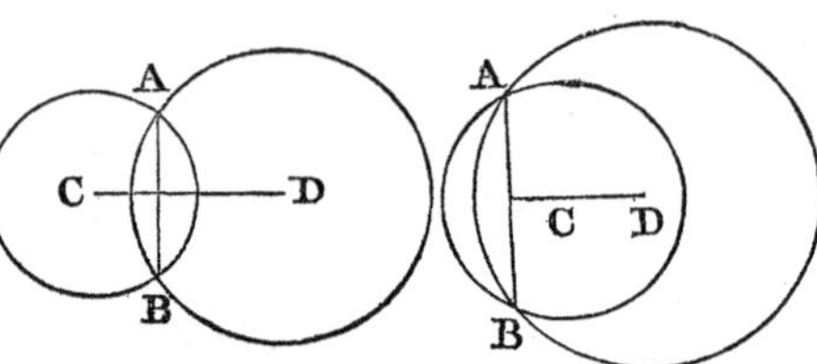

Let two circumferences cut each other in the points A and B; then will the line AB be a common chord to the two circles. Now, if a perpendicular be erected from the middle of this chord, it will pass through C and D, the centers of the two circles (Prop. VI., Schol.). But only one straight line can be drawn through two given points; therefore, the straight line which passes through the centers, will bisect the common chord at right angles.

PROPOSITION XII. THEOREM.

If two circumferences touch each other, either externally or internally, the distance of their centers must be equal to the sum or difference of their radii.

It is plain that the centers of the circles and the point of

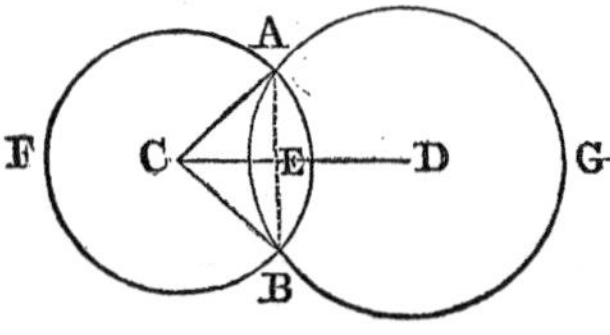

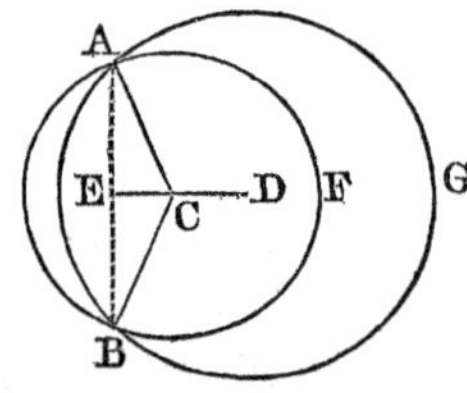

contact are in the same straight line; for, if possible, let the point of contact, A, be without the straight line CD. From A let fall upon CD, or CD produced, the perpendicular AE, and produce it to B, making BE equal to AE. Then, in the triangles ACE, BCE, the side AE is equal to EB, CE is common, and the angle AEC is equal to the angle BEC; therefore AC is equal to CB (Prop. VI., B. I.), and the point B is in the circumference ABF. In the same manner, it may be shown to be in the circumference ABG, and hence the point

B is in both circumferences. Therefore the two circumferences have two points, A and B, in common; that is, they cut each other, which is contrary to the hypothesis. Therefore, the point of contact can not be without the line joining the centers; and hence, when the circles touch each other externally, the distance of the centers CD is equal to the sum of the radii CA, DA; and when they touch internally, the distance CD is equal to the difference of the radii CA, DA. Therefore, if two circumferences, &c.

Schol. If two circumferences touch each other, externally or internally, their centers and the point of contact are in the same straight line.

PROPOSITION XIII. THEOREM.

If two circumferences cut each other, the distance between their centers is less than the sum of their radii, and greater than their difference.

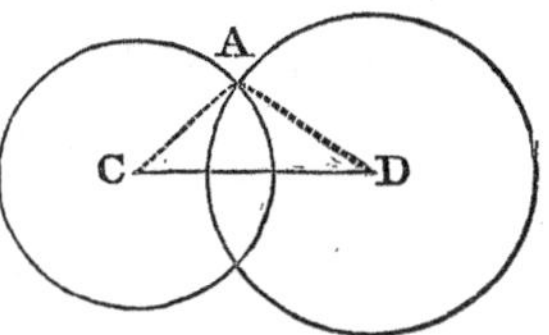

Let two circumferences cut each other in the point A. Draw the radii CA, DA; then, because any two sides of a triangle are together greater than the third side (Prop. VIII., B. I.), CD must be less than the sum of AD and AC. Also, DA must be less than the sum of CD and CA; or, subtracting CA from these unequals (Axiom 5, B. I.), CD must be greater than the difference between DA and CA. Therefore, if two circumferences, &c.

PROPOSITION XIV. THEOREM.

In equal circles, angles at the center have the same ratio with the intercepted arcs.

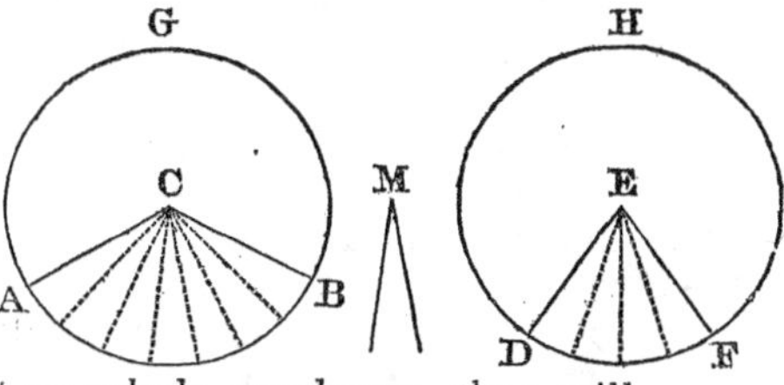

Case first. When the angles are in the ratio of two whole numbers.

Let ABG, DFH be equal circles, and let the angles ACB, DEF at their centers be in the ratio of two whole numbers; then will

the angle ACB : angle DEF : : arc AB : arc DF.

Suppose, for example, that the angles ACB, DEF are to each other as 7 to 4; or, which is the same thing, suppose that the angle M, which may serve as a common measure, is contained seven times in the angle ACB, and four times in the angle DEF. The seven partial angles into which ACB is divided, being each equal to any of the four partial angles into which DEF is divided, the partial arcs will also be equal to each other (Prop. IV.), and the entire arc AB will be to the entire arc DF as 7 to 4. Now the same reasoning would apply, if in place of 7 and 4 any whole numbers whatever were employed; therefore, if the ratio of the angles ACB, DEF can be expressed in whole numbers, the arcs AB, DF will be to each other as the angles ACB, DEF.

Case second. When the ratio of the angles can not be expressed by whole numbers.

Let ACB, ACD be two angles having any ratio whatever. Suppose ACD to be the smaller angle, and let it be placed on the greater; then will the angle ACB : angle ACD : : arc AB : arc AD.

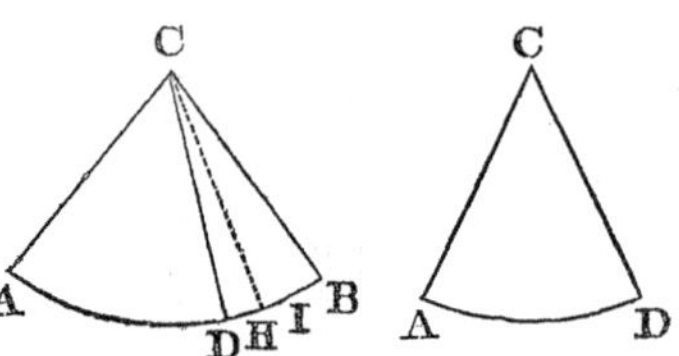

For, if this proportion is not true, the first three terms remaining the same, the fourth must be greater or less than AD. Suppose it to be greater, and that we have

Angle ACB : angle ACD : : arc AB : arc AI.

Conceive the arc AB to be divided into equal parts, each less than DI; there will be at least one point of division between D and I. Let H be that point, and join CH. The arcs AB, AH will be to each other in the ratio of two whole numbers, and, by the preceding case, we shall have

Angle ACB : angle ACH : : arc AB : arc AH.

Comparing these two proportions with each other, and observing that the antecedents are the same, we conclude that the consequents are proportional (Prop. IV., Cor., B. II.); therefore,

Angle ACD : angle ACH : : arc AI : arc AH.

But the arc AI is greater than the arc AH; therefore the angle ACD is greater than the angle ACH (Def. 2, B. II.), that is, a part is greater than the whole, which is absurd. Hence the angle ACB can not be to the angle ACD as the arc AB to an arc greater than AD.

In the same manner, it may be proved that the fourth term of the proportion can not be less than AD; therefore, it must be AD, and we have the proportion

Angle ACB · angle ACD : : arc AB : arc AD.

Cor. 1. Since the angle at the center of a circle, and the arc intercepted by its sides, are so related, that when one is increased or diminished, the other is increased or diminished in the same ratio, we may take either of these quantities as the measure of the other. Henceforth we shall take the arc AB to measure the angle ACB. It is important to observe, that in the comparison of angles, the arcs which measure them must be described with equal radii.

Cor. 2. In equal circles, *sectors are to each other as their arcs;* for sectors are equal when their angles are equal.

PROPOSITION XV. THEOREM.

An inscribed angle is measured by half the arc included between its sides.

Let BAD be an angle inscribed in the circle BAD. The angle BAD is measured by half the arc BD.

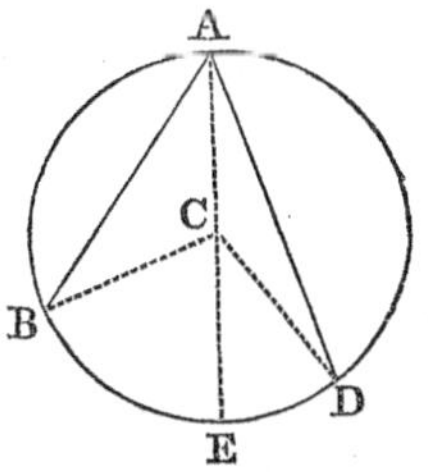

First. Let C, the center of the circle, be within the angle BAD. Draw the diameter AE, also the radii CB, CD.

Because CA is equal to CB, the angle CAB is equal to the angle CBA (Prop. X., B. I.); therefore the angles CAB, CBA are together double the angle CAB. But the angle BCE is equal (Prop. XXVII., B. I.) to the angles CAB, CBA; therefore, also, the angle BCE is double of the angle BAC. Now the angle BCE, being an angle at the center, is measured by the arc BE; hence the angle BAE is measured by the half of BE. For the same reason, the angle DAE is measured by half the arc DE. Therefore, the whole angle BAD is measured by half the arc BD.

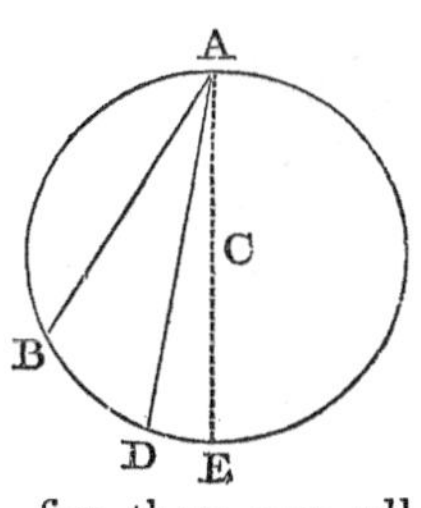

Second. Let C, the center of the circle, be without the angle BAD. Draw the diameter AE. It may be demonstrated, as in the first case, that the angle BAE is measured by half the arc BE, and the angle DAE by half the arc DE; hence their difference, BAD, is measured by half of BD. Therefore, an inscribed angle, &c.

Cor. 1. All the angles BAC, BDC, &c., inscribed in the same segment are equal, for they are all measured by half the same arc BEC. (See next fig.)

Cor. 2. Every angle inscribed in a semicircle is a right angle, because it is measured by half a semicircumference that is, the fourth part of a circumference

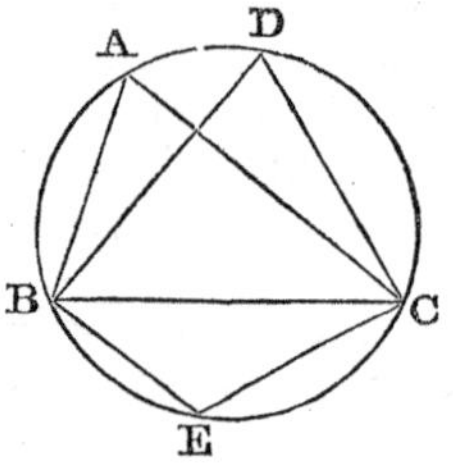

Cor. 3. Every angle inscribed in a segment greater than a semicircle is an acute angle, for it is measured by half an arc less than a semicircumference.

Every angle inscribed in a segment less than a semicircle is an obtuse angle, for it is measured by half an arc greater than a semicircumference.

Cor. 4. The opposite angles of an inscribed quadrilateral, ABEC, are together equal to two right angles; for the angle BAC is measured by half the arc BEC, and the angle BEC is measured by half the arc BAC; therefore the two angles BAC, BEC, taken together, are measured by half the circumference; hence their sum is equal to two right angles.

PROPOSITION XVI. THEOREM.

The angle formed by a tangent and a chord, is measured by half the arc included between its sides.

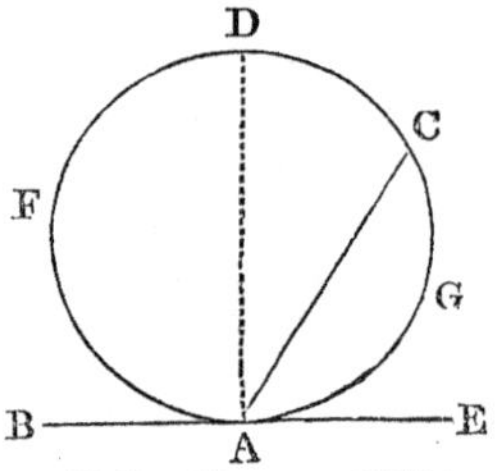

Let the straight line BE touch the circumference ACDF in the point A, and from A let the chord AC be drawn; the angle BAC is measured by half the arc AFC.

From the point A draw the diameter AD. The angle BAD is a right angle (Prop. IX.), and is measured by half the semicircumference AFD; also, the angle DAC is measured by half the arc DC (Prop. XV.); therefore, the sum of the angles BAD, DAC is measured by half the entire arc AFDC.

In the same manner, it may be shown that the angle CAE is measured by half the arc AC, included between its sides.

Cor. The angle BAC is equal to an angle inscribed in the segment AGC; and the angle EAC is equal to an angle inscribed in the segment AFC.

BOOK IV.

THE PROPORTIONS OF FIGURES

Definitions.

1. *Equal figures* are such as may be applied the one to the other, so as to coincide throughout. Thus, two circles having equal radii are equal; and two triangles, having the three sides of the one equal to the three sides of the other, each to each, are also equal.

2. *Equivalent figures* are such as contain equal areas Two figures may be equivalent, however dissimilar. Thus, a circle may be equivalent to a square, a triangle to a rectangle, &c.

3. *Similar figures* are such as have the angles of the one equal to the angles of the other, each to each, and the sides about the equal angles proportional. Sides which have the same position in the two figures, or which are adjacent to equal angles, are called *homologous.* The equal angles may also be called homologous angles.

Equal figures are always similar, but similar figures may be very unequal.

4. Two sides of one figure are said to be *reciprocally proportional* to two sides of another, when one side of the first is to one side of the second, as the remaining side of the second is to the remaining side of the first.

5. In different circles, *similar arcs, sectors, or segments,* are those which correspond to equal angles at the center.

Thus, if the angles A and D are equal, the arc BC will be similar to the arc EF, the sector ABC to the sector DEF, and the segment BGC to the segment EHF.

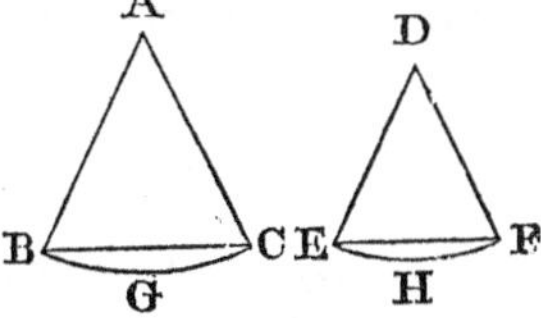

6. The *altitude of a triangle* is the perpendicular let fall from the vertex of an angle on the opposite side, taken as a base, or on the base produced.

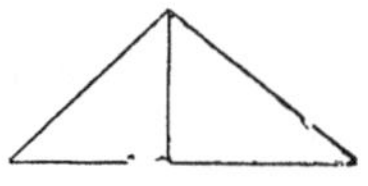

7. The *altitude of a parallelogram* is the perpendicular drawn to the base from the opposite side.

8. The *altitude of a trapezoid* is the distance between its parallel sides.

PROPOSITION I. THEOREM.

Parallelograms which have equal bases and equal altitudes are equivalent.

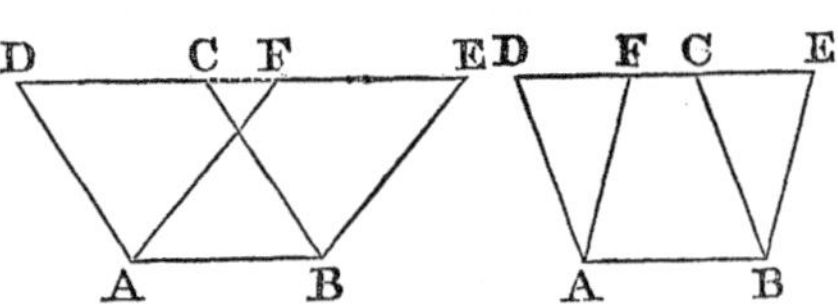

Let the parallelograms ABCD, ABEF be placed so that their equal bases shall coincide with each other. Let AB be the common base; and, since the two parallelograms are supposed to have the same altitude, their upper bases, DC, FE, will be in the same straight line parallel to AB.

Now, because ABCD is a parallelogram, DC is equal to AB (Prop. XXIX., B. I.). For the same reason, FE is equal to AB, wherefore DC is equal to FE; hence, if DC and FE be taken away from the same line DE, the remainders CE and DF will be equal. But AD is also equal to BC, and AF to BE; therefore the triangles DAF, CBE are mutually equilateral, and consequently equal.

Now if from the quadrilateral ABED we take the triangle ADF, there will remain the parallelogram ABEF; and if from the same quadrilateral we take the triangle BCE, there will remain the parallelogram ABCD. Therefore, the two parallelograms ABCD, ABEF, which have the same base and the same altitude, are equivalent.

Cor. Every parallelogram is equivalent to the rectangle which has the same base and the same altitude.

PROPOSITION II. THEOREM.

Every triangle is half of the parallelogram which has the same base and the same altitude.

Let the parallelogram ABDE and the triangle ABC have the same base, AB, and the same altitude; the triangle is half of the parallelogram.

Complete the parallelogram ABFC; then the parallelogram ABFC is equivalent to the parallelogram ABDE, because they have the same base and the same altitude (Prop. I.). But the triangle ABC is half of the parallelogram ABFC (Prop. XXIX., Cor., B. I.); wherefore the triangle ABC is also half of the parallelogram ABDE. Therefore, every triangle, &c.

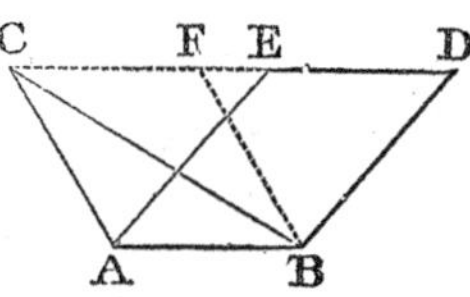

Cor. 1. Every triangle is half of the rectangle which has the same base and altitude.

Cor. 2. Triangles which have equal bases and equal altitudes are equivalent.

PROPOSITION III. THEOREM.

Two rectangles of the same altitude, are to each other as their bases.

Let ABCD, AEFD be two rectangles which have the common altitude AD; they are to each other as their bases AB, AE.

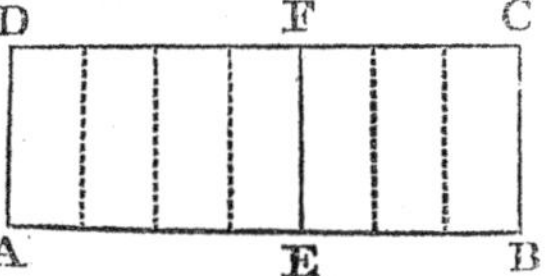

Case first. When the bases are in the ratio of two whole numbers, for example, as 7 to 4. If AB be divided into seven equal parts, AE will contain four of those parts. At each point of division, erect a perpendicular to the base; seven partial rectangles will thus be formed, all equal to each other, since they have equal bases and altitudes (Prop. I.). The rectangle ABCD will contain seven partial rectangles, while AEFD will contain four; therefore the rectangle ABCD is to the rectangle AEFD as 7 to 4, or as AB to AE. The same reasoning is applicable to any other ratio than that of 7 to 4, therefore, whenever the ratio of the bases can be expressed in whole numbers, we shall have

ABCD : AEFD : : AB : AE.

Case second. When the ratio of the bases can not be expressed in whole numbers, it is still true that

ABCD : AEFD : : AB : AE.

For, if this proportion is not true, the first three terms remaining the same, the fourth must be greater or less than AE. Suppose it to be greater, and that we have

ABCD : AEFD : : AB : AG.

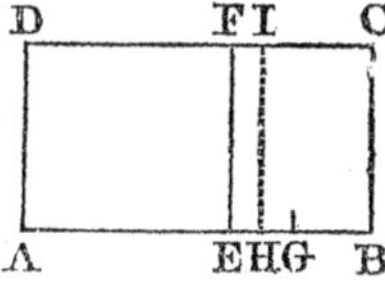

Conceive the line AB to be divided into

equal parts, each less than EG; there will be at least one point of division between E and G. Let H be that point, and draw the perpendicular HI. The bases AB, AH will be to each other in the ratio of two whole numbers, and by the preceding case we shall have

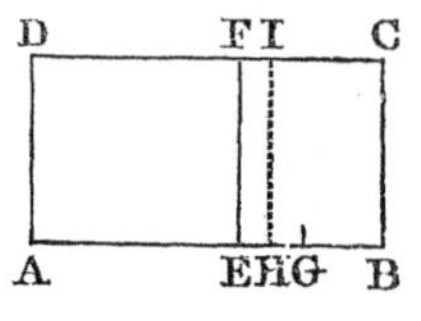

ABCD : AHID : : AB : AH.

But, by hypothesis, we have

ABCD : AEFD : : AB : AG.

In these two proportions the antecedents are equal; therefore the consequents are proportional (Prop. IV., Cor., B. II.), and we have

AHID : AEFD : : AH : AG.

But AG is greater than AH; therefore the rectangle AEFD is greater than AHID (Def. 2, B. II.); that is, a part is greater than the whole, which is absurd. Therefore ABCD can not be to AEFD as AB to a line greater than AE.

In the same manner, it may be shown that the fourth term of the proportion can not be less than AE; hence it must be AE, and we have the proportion

ABCD : AEFD : : AB : AE.

Therefore, two rectangles, &c.

PROPOSITION IV. THEOREM.

Any two rectangles are to each other as the products of their bases by their altitudes.

Let ABCD, AEGF be two rectangles; the ratio of the rectangle ABCD to the rectangle AEGF, is the same with the ratio of the product of AB by AD, to the product of AE by AF; that is,

ABCD : AEGF : : AB × AD : AE × AF.

Having placed the two rectangles so that the angles at A are vertical, produce the sides GE, CD till they meet in H. The two rectangles ABCD, AEHD have the same altitude AD; they are, therefore, as their bases AB, AE (Prop. III.). So, also, the rectangles AEHD, AEGF, having the same altitude AE, are to each other as their bases AD, AF. Thus, we have the two proportions

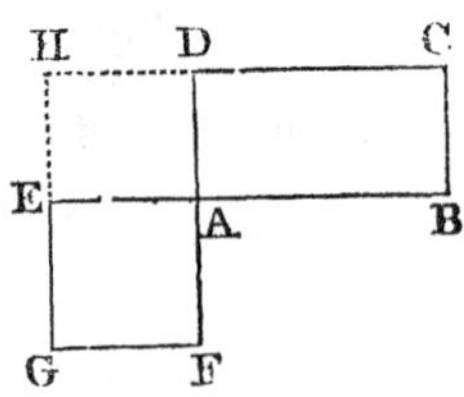

ABCD : AEHD : : AB : AE,
AEHD : AEGF : : AD : AF.

Hence (Prop. XI., Cor., B. II.),

ABCD : AEGF : : AB×AD : AE×AF.

Scholium. Hence we may take as the *measure* of a rectangle the product of its base by its altitude; provided we understand by it the product of two numbers, one of which is the number of linear units contained in the base, and the other the number of linear units contained in the altitude.

PROPOSITION V. THEOREM.

The area of a parallelogram is equal to the product of its base by its altitude.

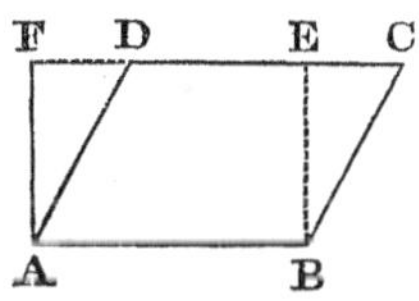

Let ABCD be a parallelogram, AF its altitude, and AB its base; then is its surface measured by the product of AB by AF. For, upon the base AB, construct a rectangle having the altitude AF; the parallelogram ABCD is equivalent to the rectangle ABEF (Prop. I., Cor.). But the rectangle ABEF is measured by AB×AF (Prop. IV., Schol.); therefore the area of the parallelogram ABCD is equal to AB×AF.

Cor. Parallelograms of the same base are to each other as their altitudes, and parallelograms of the same altitude are to each other as their bases; for magnitudes have the same ratio that their equimultiples have (Prop. VIII., B. II.).

PROPOSITION VI. THEOREM.

The area of a triangle is equal to half the product of its base by its altitude.

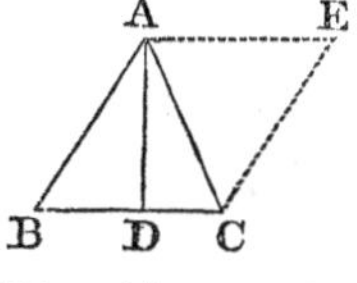

Let ABC be any triangle, BC its base, and AD its altitude; the area of the triangle ABC is measured by half the product of BC by AD.

For, complete the parallelogram ABCE. The triangle ABC is half of the parallelogram ABCE (Prop. II.); but the area of the parallelogram is equal to BC×AD (Prop. V.); hence the area of the triangle is equal to one half of the product of BC by AD. Therefore, the area of a triangle, &c.

Cor. 1. Triangles of the same altitude are to each other as their bases, and triangles of the same base are to each other as their altitudes.

Cor 2 Equivalent triangles, whose bases are equal, have

equal altitudes; and equivalent triangles, whose altitudes are equal, have equal bases.

PROPOSITION VII. THEOREM.

The area of a trapezoid is equal to half the product of its altitude by the sum of its parallel sides.

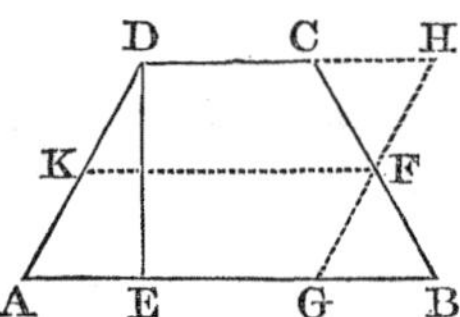

Let ABCD be a trapezoid, DE its altitude, AB and CD its parallel sides; its area is measured by half the product of DE, by the sum of its sides AB, CD.

Bisect BC in F, and through F draw GH parallel to AD, and produce DC to H. In the two triangles BFG, CFH, the side BF is equal to CF by construction, the vertical angles BFG, CFH are equal (Prop. V., B. I.), and the angle FCH is equal to the alternate angle FBG, because CH and BG are parallel (Prop. XXIII., B. I.); therefore the triangle CFH is equal to the triangle BFG. Now, if from the whole figure, ABFHD, we take away the triangle CFH, there will remain the trapezoid ABCD; and if from the same figure, ABFHD, we take away the equal triangle BFG, there will remain the parallelogram AGHD. Therefore the trapezoid ABCD is equivalent to the parallelogram AGHD, and is measured by the product of AG by DE.

Also, because AG is equal to DH, and BG to CH, therefore the sum of AB and CD is equal to the sum of AG and DH, or twice AG. Hence AG is equal to half the sum of the parallel sides AB, CD; therefore the area of the trapezoid ABCD is equal to half the product of the altitude DE by the sum of the bases AB, CD.

Cor. If through the point F, the middle of BC, we draw FK parallel to the base AB, the point K will also be the middle of AD. For the figure AKFG is a parallelogram, as also DKFH, the opposite sides being parallel. Therefore AK is equal to FG, and DK to HF. But FG is equal to FH, since the triangles BFG, CFH are equal; therefore AK is equal to DK.

Now, since KF is equal to AG, the area of the trapezoid is equal to DE × KF. Hence *the area of a trapezoid is equal to its altitude, multiplied by the line which joins the middle points of the sides which are not parallel.*

PROPOSITION VIII. THEOREM.

If a straight line is divided into any two parts, the square of the whole line is equivalent to the squares of the two parts, together with twice the rectangle contained by the parts.

Let the straight line AB be divided into any two parts in C; the square on AB is equivalent to the squares on AC CB, together with twice the rectangle contained by AC, CB; that is,

$$AB^2, \text{ or } (AC+CB)^2=AC^2+CB^2+2AC\times CB.$$

Upon AB describe the square ABDE; take AF equal to AC, through F draw FG parallel to AB, and through C draw CH parallel to AE.

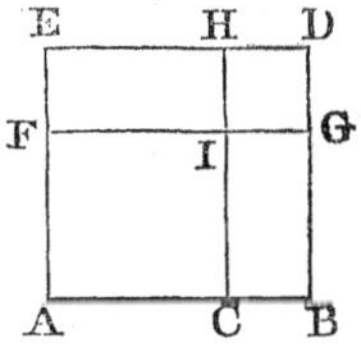

E H D F I G A C B

The square ABDE is divided into four parts: the first, ACIF, is the square on AC, since AF was taken equal to AC. The second part, IGDH, is the square on CB; for, because AB is equal to AE, and AC to AF, therefore BC is equal to EF (Axiom 3, B. I.). But, because BCIG is a parallelogram, GI is equal to BC; and because DEFG is a parallelogram, DG is equal to EF (Prop. XXIX., B. I.); therefore HIGD is equal to a square described on BC. If these two parts are taken from the entire square, there will remain the two rectangles BCIG, EFIH, each of which is measured by AC×CB; therefore the whole square on AB is equivalent to the squares on AC and CB, together with twice the rectangle of AC×CB. Therefore, if a straight line, &c.

Cor. The square of any line is equivalent to four times the square of half that line. For, if AC is equal to CB, the four figures AI, CG, FH, ID become equal squares.

Scholium. This proposition is expressed algebraically thus:

$$(a+b)^2=a^2+2ab+b^2.$$

PROPOSITION IX. THEOREM.

The square described on the difference of two lines, is equivalent to the sum of the squares of the lines, diminished by twice the rectangle contained by the lines.

Let AB, BC be any two lines, and AC their difference: the square described on AC is equivalent to the sum of the squares on AB and CB, diminished by twice the rectangle contained by AB, CB; that is,

$$AC^2, \text{ or } (AB-BC)^2=AB^2+BC^2-2AB\times BC.$$

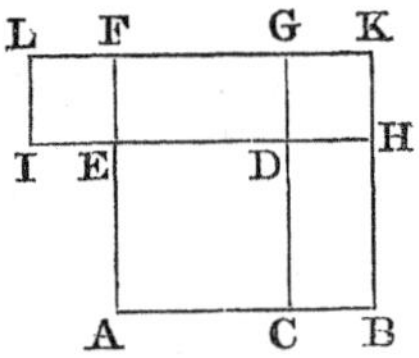

Upon AB describe the square ABKF; take AE equal to AC, through C draw CG parallel to BK, and through E draw HI parallel to AB, and complete the square EFLI.

Because AB is equal to AF, and AC to AE; therefore CB is equal to EF, and GK to LF. Therefore LG is equal to FK or AB; and hence the two rectangles CBKG, GLID are each measured by AB× BC. If these rectangles are taken from the entire figure ABKLIE, which is equivalent to AB^2+BC^2, there will evidently remain the square ACDE. Therefore, the square described, &c.

Scholium. This proposition is expressed algebraically thus:

$$(a-b)^2=a^2-2ab+b^2.$$

Cor.

$$(a+b)^2-(a-b)^2=4ab.$$

PROPOSITION X. THEOREM.

The rectangle contained by the sum and difference of two lines, is equivalent to the difference of the squares of those lines

Let AB, BC be any two lines; the rectangle contained by the sum and difference of AB and BC, is equivalent to the difference of the squares on AB and BC; that is,

$$(AB+BC)\times(AB-BC)=AB^2-BC^2.$$

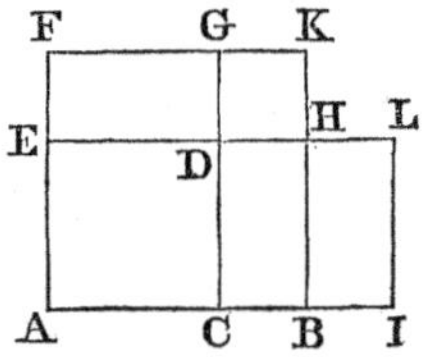

Upon AB describe the square ABKF, and upon AC describe the square ACDE; produce AB so that BI shall be equal to BC, and complete the rectangle AILE.

The base AI of the rectangle AILE is the sum of the two lines AB, BC, and its altitude AE is the difference of the same lines; therefore AILE is the rectangle contained by the sum and difference of the lines AB, BC. But this rectangle is composed of the two parts ABHE and BILH; and the part BILH is equal to the rectangle EDGF, for BH is equal to DE, and BI is equal to EF. Therefore AILE is equivalent to the figure ABHDGF. But ABHDGF is the excess of the square ABKF above the square DHKG, which is the square of BC; therefore,

$$(AB+BC)\times(AB-BC)=AB^2-BC^2.$$

Scholium. This proposition is expressed algebraically thus:

$$(a+b)\times(a-b)=a^2-b^2.$$

PROPOSITION XI. THEOREM.

In any right-angled triangle, the square described on the hypothenuse is equivalent to the sum of the squares on the other two sides.

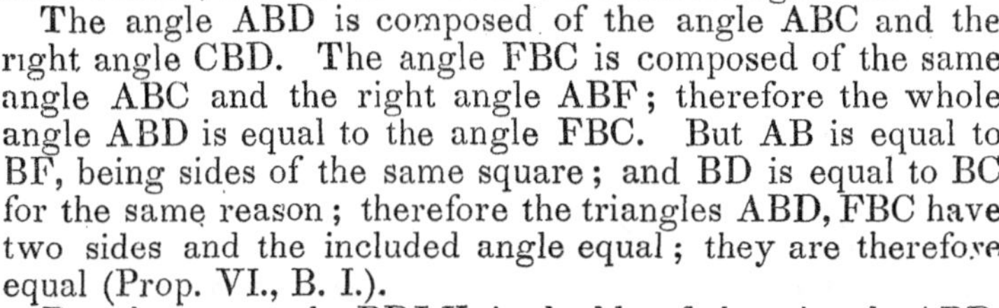

Let ABC be a right-angled triangle, having the right angle BAC; the square described upon the side BC is equivalent to the sum of the squares upon BA, AC.

On BC describe the square BCED, and on BA, AC the squares BG, CH; and through A draw AL parallel to BD, and join AD, FC.

Then, because each of the angles BAC, BAG is a right angle, CA is in the same straight line with AG (Prop. III., B. I.). For the same reason, BA and AH are in the same straight line.

The angle ABD is composed of the angle ABC and the right angle CBD. The angle FBC is composed of the same angle ABC and the right angle ABF; therefore the whole angle ABD is equal to the angle FBC. But AB is equal to BF, being sides of the same square; and BD is equal to BC for the same reason; therefore the triangles ABD, FBC have two sides and the included angle equal; they are therefore equal (Prop. VI., B. I.).

But the rectangle BDLK is double of the triangle ABD, because they have the same base, BD, and the same altitude, BK (Prop. II., Cor. 1); and the square AF is double of the triangle FBC, for they have the same base, BF, and the same altitude, AB. Now the doubles of equals are equal to one another (Axiom 6, B. I.); therefore the rectangle BDLK is equivalent to the square AF.

In the same manner, it may be demonstrated that the rectangle CELK is equivalent to the square AI; therefore the whole square BCED, described on the hypothenuse, is equivalent to the two squares ABFG, ACIH, described on the two other sides; that is,

$$BC^2=AB^2+AC^2.$$

Cor. 1. The square of one of the sides of a right-angled

triangle, is equivalent to the square of the hypothenuse, diminished by the square of the other side; that is,

$$AB^2 = BC^2 - AC^2.$$

Cor. 2. The square BCED, and the rectangle BKLD, having the same altitude, are to each other as their bases BC, BK (Prop. III.). But the rectangle BKLD is equivalent to the square AF; therefore,

$$BC^2 : AB^2 :: BC : BK.$$

In the same manner,

$$BC^2 : AC^2 :: BC : KC.$$

Therefore (Prop. IV., Cor., B. II.),

$$AB^2 : AC^2 :: BK : KC.$$

That is, in any right-angled triangle, if a line be drawn from the right angle perpendicular to the hypothenuse, *the squares of the two sides are proportional to the adjacent segments of the hypothenuse;* also, *the square of the hypothenuse is to the square of either of the sides, as the hypothenuse is to the segment adjacent to that side.*

Cor. 3. Let ABCD be a square, and AC its diagonal; the triangle ABC being right-angled and isosceles, we have

$$AC^2 = AB^2 + BC^2 = 2AB^2;$$

therefore *the square described on the diagonal of a square, is double of the square described on a side.*

If we extract the square root of each member of this equation, we shall have

$$AC = AB\sqrt{2}; \text{ or } AC : AB :: \sqrt{2} : 1.$$

PROPOSITION XII. THEOREM.

In any triangle, the square of a side opposite an acute angle, is less than the squares of the base and of the other side, by twice the rectangle contained by the base, and the distance from the acute angle to the foot of the perpendicular let fall from the opposite angle.

Let ABC be any triangle, and the angle at C one of its acute angles, and upon BC let fall the perpendicular AD from the opposite angle; then will

$$AB^2 = BC^2 + AC^2 - 2BC \times CD.$$

First. When the perpendicular falls within the triangle ABC, we have $BD = BC - CD$, and therefore $BD^2 = BC^2 + CD^2 - 2BC \times CD$ (Prop. IX.). To each of these equals add AD^2; then $BD^2 + AD^2 = BC^2 + CD^2 + AD^2 - 2BC \times CD$. But in the right-angled triangle

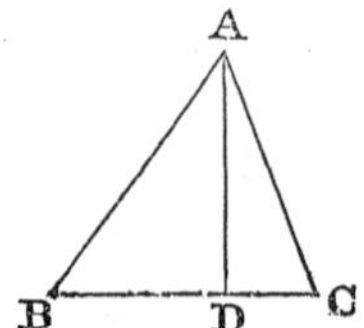

ABD, $BD^2+AD^2=AB^2$; and in the triangle ADC, $CD^2+AD^2=AC^2$ (Prop. XI.) ; therefore

$$AB^2=BC^2+AC^2-2BC\times CD.$$

Secondly. When the perpendicular falls without the triangle ABC, we have $BD=CD-BC$, and therefore $BD^2=CD^2+BC^2-2CD\times BC$ (Prop. IX.). To each of these equals add AD^2 ; then $BD^2+AD^2=CD^2+AD^2+BC^2-2CD\times BC$. But $BD^2+AD^2=AB^2$; and $CD^2+AD^2=AC^2$; therefore

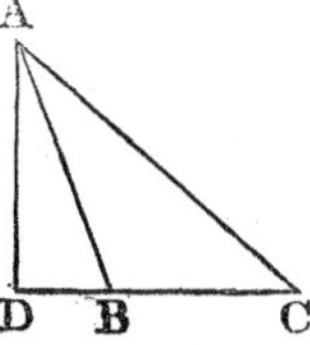

$$AB^2=BC^2+AC^2-2BC\times CD.$$

Scholium. When the perpendicular AD falls upon AB, this proposition reduces to the same as Prop. XI., Cor. 1.

PROPOSITION XIII. THEOREM.

In obtuse-angled triangles, the square of the side opposite the obtuse angle, is greater than the squares of the base and the other side, by twice the rectangle contained by the base, and the distance from the obtuse angle to the foot of the perpendicular let fall from the opposite angle on the base produced.

Let ABC be an obtuse-angled triangle, having the obtuse angle ABC, and from the point A let AD be drawn perpendicular to BC produced; the square of AC is greater than the squares of AB, BC by twice the rectangle $BC\times BD$.

For CD is equal to $BC+BD$; therefore $CD^2=BC^2+BD^2+2BC\times BD$ (Prop. VIII.). To each of these equals add AD^2; then $CD^2+AD^2=BC^2+BD^2+AD^2+2BC\times BD$. But AC^2 is equal to CD^2+AD^2 (Prop. XI.), and AB^2 is equal to BD^2+AD^2; therefore $AC^2=BC^2+AB^2+2BC\times BD$. Therefore, in obtuse-angled triangles, &c.

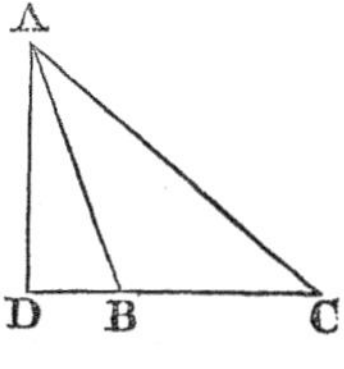

Scholium. The right-angled triangle is the only one in which the sum of the squares of two sides is equivalent to the square on the third side; for, if the angle contained by the two sides is acute, the sum of their squares is greater than the square of the opposite side; if obtuse, it is less.

PROPOSITION XIV. THEOREM.

In any triangle, if a straight line is drawn from the vertex to the middle of the base, the sum of the squares of the other two sides is equivalent to twice the square of the bisecting line, together with twice the square of half the base.

Let ABC be a triangle having a line AD drawn from the

middle of the base to the opposite angle; the squares of BA and AC are together double of the squares of AD and BD.

From A draw AE perpendicular to BC; then, in the triangle ABD, by Prop. XIII.,

$$AB^2=AD^2+DB^2+2DB\times DE;$$

and, in the triangle ADC, by Prop. XII.,

$$AC^2=AD^2+DC^2-2DC\times DE.$$

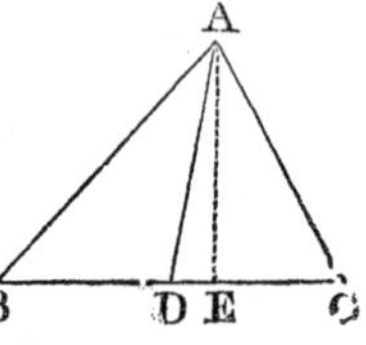

Hence, by adding these equals, and observing that BD=DC, and therefore $BD^2=DC^2$, and $DB\times DE=DC\times DE$, we obtain

$$AB^2+AC^2=2AD^2+2DB^2.$$

Therefore, in any triangle, &c.

PROPOSITION XV. THEOREM.

In every parallelogram the squares of the sides are together equivalent to the squares of the diagonals.

Let ABCD be a parallelogram, of which the diagonals are AC and BD; the sum of the squares of AC and BD is equivalent to the sum of the squares of AB, BC, CD, DA.

The diagonals AC and BD bisect each other in E (Prop. XXXII., B. I.); therefore, in the triangle ABD (Prop. XIV.),

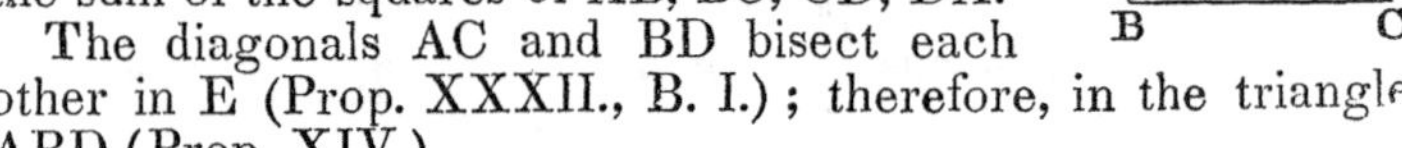

$$AB^2+AD^2=2BE^2+2AE^2;$$

and, in the triangle BDC,

$$CD^2+BC^2=2BE^2+2EC^2.$$

Adding these equals, and observing that AE is equal to EC, we have

$$AB^2+BC^2+CD^2+AD^2=4BE^2+4AE^2.$$

But $4BE^2=BD^2$, and $4AE^2=AC^2$ (Prop. VIII., Cor.); therefore

$$AB^2+BC^2+CD^2+AD^2=BD^2+AC^2.$$

Therefore, in every parallelogram, &c.

PROPOSITION XVI. THEOREM.

If a straight line be drawn parallel to the base of a triangle, it will cut the other sides proportionally; and if the sides be cut proportionally, the cutting line will be parallel to the base of the triangle.

Let DE be drawn parallel to BC, the base of the triangle ABC then will AD DB : : AE : EC.

Join BE and DC; then the triangle BDE is equivalent to the triangle DEC, because they have the same base, DE, and the same altitude, since their vertices B and C are in a line parallel to the base (Prop. II., Cor. 2).

The triangles ADE, BDE, whose common vertex is E, having the same altitude, are to each other as their bases AD, DB (Prop. VI., Cor. 1); hence

ADE : BDE : : AD : DB.

The triangles ADE, DEC, whose common vertex is D, having the same altitude, are to each other as their bases. AE, EC; therefore

ADE : DEC : : AE : EC.

But, since the triangle BDE is equivalent to the triangle DEC, therefore (Prop. IV., B. II.),

AD : DB : : AE : EC.

Conversely, let DE cut the sides AB, AC, so that AD : DB : : AE : EC; then DE will be parallel to BC.

For AD : DB : : ADE : BDE (Prop. VI., Cor. 1); and AE : EC : : ADE : DEC; therefore (Prop. IV., B. II.), ADE : BDE : : ADE : DEC; that is, the triangles BDE, DEC have the same ratio to the triangle ADE; consequently, the triangles BDE, DEC are equivalent, and having the same base DE, their altitudes are equal (Prop. VI., Cor. 2), that is, they are between the same parallels. Therefore, if a straight line, &c.

Cor. 1. Since, by this proposition, AD : DB : : AE : EC; by composition, AD+DB : AD : : AE+EC : AE (Prop. VI., B. II.), or AB : AD : : AC : AE; also, AB : BD : : AC : EC.

Cor. 2. *If two lines be drawn parallel to the base of a triangle, they will divide the other sides proportionally.* For, because FG is drawn parallel to BC, by the preceding proposition, AF : FB : : AG : GC. Also, by the last corollary, because DE is parallel to FG, AF : DF : : AG : EG. Therefore DF : FB : : EG : GC (Prop. IV., Cor., B. II.). Also, AD : DF : : AE : EG.

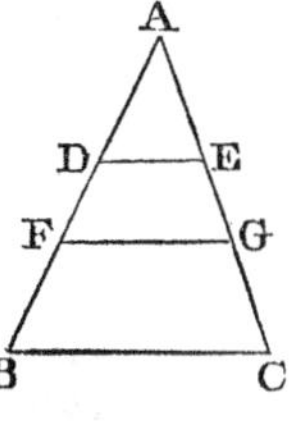

Cor. 3. If any number of lines be drawn parallel to the base of a triangle, the sides will be cut proportionally.

PROPOSITION XVII. THEOREM.

The line which bisects the vertical angle of a triangle, divides the base into two segments, which are proportional to the adjacent sides.

Let the angle BAC of the triangle ABC be bisected by the straight line AD; then will

BD : DC : : BA : AC.

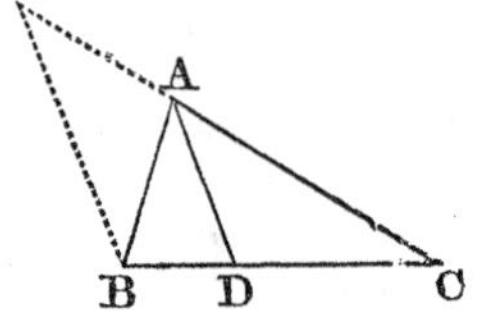

Through the point B draw BE parallel to DA, meeting CA produced in E. The triangle ABE is isosceles. For, since AD is parallel to EB, the angle ABE is equal to the alternate angle DAB (Prop. XXIII., B. I.); and the exterior angle CAD is equal to the interior and opposite angle AEB. But, by hypothesis, the angle DAB is equal to the angle DAC; therefore the angle ABE is equal to AEB, and the side AE to the side AB (Prop. XI., B. I.).

And because AD is drawn parallel to BE, the base of the triangle BCE (Prop. XVI.),

BD : DC : : EA : AC.

But AE is equal to AB, therefore

BD : DC : : BA : AC.

Therefore, the line, &c.

Scholium. The line which bisects the *exterior* angle of a triangle, divides the base produced into segments, which are proportional to the adjacent sides.

Let the line AD bisect the exterior angle CAE of the triangle ABC; then

BD : DC : : BA : AC.

Through C draw CF parallel to AD; then it may be proved, as in the preceding proposition, that the angle ACF is equal to the angle AFC, and AF equal to AC. And because FC is parallel to AD (Prop. XVI., Cor. 1), BD : DC : BA : AF. But AF is equal to AC; therefore

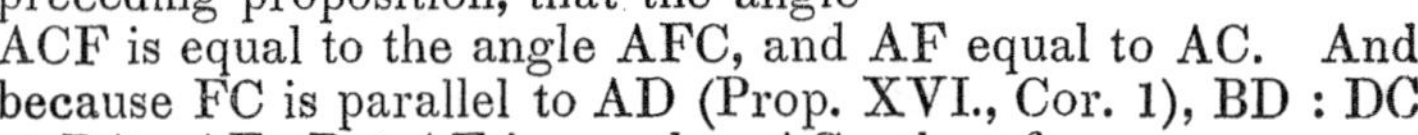

BD : DC : : BA : AC.

PROPOSITION XVIII. THEOREM.

Equiangular triangles have their homologous sides proportional, and are similar.

Let ABC, DCE be two equiangular triangles, having the angle BAC equal to the angle CDE, and the angle ABC equal to the angle DCE, and, consequently, the angle ACB equal to the angle DEC; then the homologous sides will be proportional, and we shall have

BC : CE : : BA : CD : : AC : DE.

Place the triangle DCE so that the side CE may be contiguous to BC, and in the same straight line with it; and produce the sides BA, ED till they meet in F.

Because BCE is a straight line, and the angle ACB is equal to the angle DEC, AC is parallel to EF (Prop. XXII., B. I.). Again, because the angle ABC is equal to the angle DCE, the line AB is parallel to DC; therefore the figure ACDF is a parallelogram, and, consequently, AF is equal to CD, and AC to FD (Prop. XXIX., B. I.).

And because AC is parallel to FE, one of the sides of the triangle FBE, BC : CE : : BA : AF (Prop. XVI.); but AF is equal to CD; therefore

BC : CE : : BA : CD.

Again, because CD is parallel to BF, BC : CE : : FD : DE But FD is equal to AC; therefore

BC : CE : : AC : DE.

And, since these two proportions contain the same ratio BC : CE, we conclude (Prop. IV., B. II.)

BA : CD : : AC : DE.

Therefore the equiangular triangles ABC, DCE have their homologous sides proportional; hence, by Def. 3, they are similar.

Cor. Two triangles are similar when they have two angles equal, each to each, for then the third angles must also be equal.

Scholium. In similar triangles the homologous sides are opposite to the equal angles; thus, the angle ACB being equal to the angle DEC, the side AB is homologous to DC, and so with the other sides.

PROPOSITION XIX. THEOREM.

Two triangles which have their homologous sides proportional, are equiangular and similar.

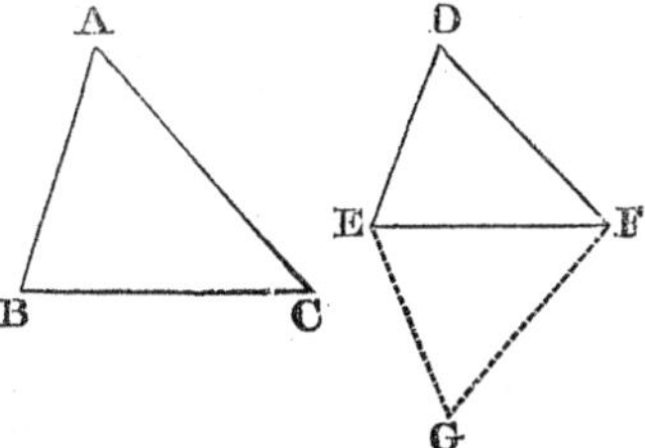

Let the triangles ABC, DEF have their sides proportional, so that BC : EF : : AB : DE : : AC : DF; then will the triangles have their angles equal, viz.: the angle A equal to the angle D, B equal to E, and C equal to F.

At the point E, in the straight line EF, make the angle FEG equal to B, and at the point F make the angle EFG equal to C; the third angle G will be equal to the third angle A, and the two triangles ABC, GEF will be equiangular (Prop. XXVII., Cor. 2, B. I.); therefore by the preceding theorem,

BC : EF : : AB : GE.

But, by hypothesis,

BC : EF : : AB : DE;

therefore GE is equal to DE.

Also, by the preceding theorem,

BC : EF : : AC : GF;

but, by hypothesis,

BC : EF : : AC : DF;

consequently, GF is equal to DF. Therefore the triangles GEF, DEF have their three sides equal, each to each; hence their angles also are equal (Prop. XV., B. I.). But, by construction, the triangle GEF is equiangular to the triangle ABC; therefore, also, the triangles DEF, ABC are equiangular and similar. Wherefore, two triangles, &c.

PROPOSITION XX. THEOREM.

Two triangles are similar, when they have an angle of the one equal to an angle of the other, and the sides containing those angles proportional.

Let the triangles ABC, DEF have the angle A of the one, equal to the angle D of the other, and let AB : DE : : AC : DF; the triangle ABC is similar to the triangle DEF.

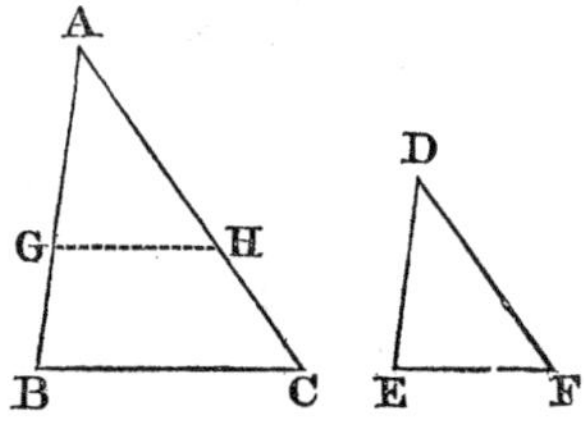

Take AG equal to DE, also AH equal to DF, and join GH. Then the triangles AGH, DEF are equal, since two sides and the included angle in the one, are respectively equal to two sides and the included angle in the other (Prop. VI., B. I.). But, by hypothesis, AB : DE : : AC : DF; therefore

AB : AG : : AC : AH;

that is, the sides AB, AC, of the triangle ABC, are cut proportionally by the line GH; therefore GH is parallel to BC (Prop. XVI.). Hence (Prop. XXIII., B. I.) the angle AGH is equal to ABC, and the triangle AGH is similar to the triangle ABC. But the triangle DEF has been shown to be equal to the triangle AGH; hence the triangle DEF is similar to the triangle ABC. Therefore, two triangles, &c.

PROPOSITION XXI. THEOREM.

Two triangles are similar, when they have their homologous sides parallel or perpendicular to each other.

Let the triangles ABC, *abc*, DEF have their homologous sides parallel or perpendicular to each other; the triangles are similar.

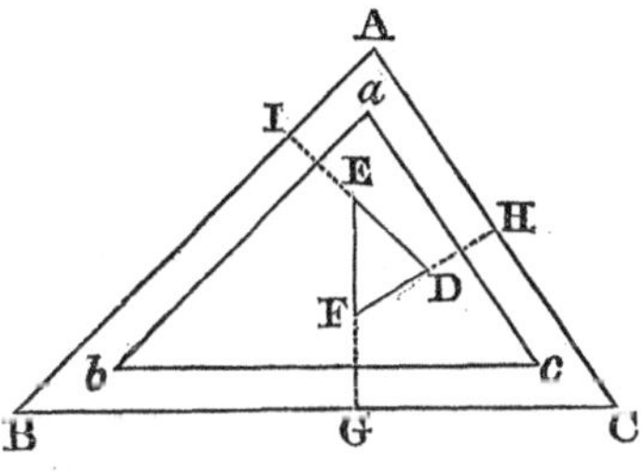

First. Let the homologous sides be parallel to each other. If the side AB is parallel to *ab*, and BC to *bc*, the angle B is equal to the angle *b* (Prop. XXVI., B. I.); also, if AC is parallel to *ac*, the angle C is equal to the angle *c*; and hence the angle A is equal to the angle *a*. Therefore the triangles ABC, *abc* are equiangular, and consequently similar.

Secondly. Let the homologous sides be perpendicular to each other. Let the side DE be perpendicular to AB, and the side DF to AC. Produce DE to I, and DF to H; then, in the quadrilateral AIDH, the two angles I and H are right angles. But the four angles of a quadrilateral are together equal to four right angles (Prop. XXVIII., Cor. 1, B. I.); therefore the two remaining angles IAH, IDH are together equal to two right angles. But the two angles EDF, IDH are together equal to two right angles (Prop. II., B. I.); therefore the angle EDF is equal to IAH or BAC.

In the same manner, if the side EF is also perpendicular to BC, it may be proved that the angle DFE is equal to C, and, consequently, the angle DEF is equal to B; hence the triangles ABC, DEF are equiangular and similar. Therefore, two triangles &c.

Scholium. When the sides of the two triangles are parallel, the parallel sides are homologous; but when the sides are perpendicular to each other, the perpendicular sides are homologous. Thus DE is homologous to AB, DF to AC, and EF to BC

PROPOSITION XXII. THEOREM.

In a right-angled triangle, if a perpendicular is drawn from the right angle to the hypothenuse,

1st. *The triangles on each side of the perpendicular are similar to the whole triangle and to each other.*

2d. *The perpendicular is a mean proportional between the segments of the hypothenuse.*

3d. *Each of the sides is a mean proportional between the hypothenuse and its segment adjacent to that side.*

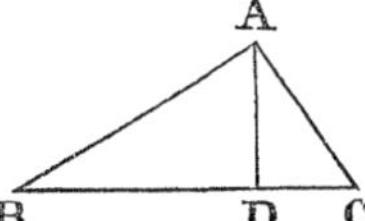

Let ABC be a right-angled triangle, having the right angle BAC, and from the angle A let AD be drawn perpendicular to the hypothenuse BC.

First. The triangles ABD, ACD are similar to the whole triangle ABC, and to each other.

The triangles BAD, BAC have the common angle B, also the angle BAC equal to BDA, each of them being a right angle, and, therefore, the remaining angle ACB is equal to the remaining angle BAD (Prop. XXVII., Cor. 2, B. I.); therefore the triangles ABC, ABD are equiangular and similar. In like manner, it may be proved that the triangle ADC is equiangular and similar to the triangle ABC; therefore the three triangles ABC, ABD, ACD are equiangular and similar to each other.

Secondly. The perpendicular AD is a mean proportional between the segments BD, DC of the hypothenuse. For, since the triangle ABD is similar to the triangle ADC, their homologous sides are proportional (Def. 3), and we have

BD : AD : : AD : DC.

Thirdly. Each of the sides AB, AC is a mean proportional between the hypothenuse and the segment adjacent to that side. For, since the triangle BAD is similar to the triangle BAC, we have

BC : BA : : BA : BD.

And, since the triangle ABC is similar to the triangle ACD we have

BC : CA : : CA : CD.

Therefore, in a right-angled triangle, &c.

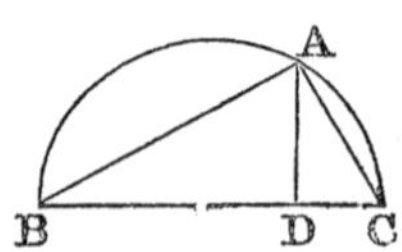

Cor. If from a point A, in the circumference of a circle, two chords AB, AC are drawn to the extremities of the diameter BC, the triangle BAC will be right-angled at A (Prop. XV., Cor. 2, B. III.); therefore

the perpendicular AD is a mean proportional between BD and DC, the two segments of the diameter; that is,

$$AD^2 = BD \times DC.$$

PROPOSITION XXIII. THEOREM.

Two triangles, having an angle in the one equal to an angle in the other, are to each other as the rectangles of the sides which contain the equal angles.

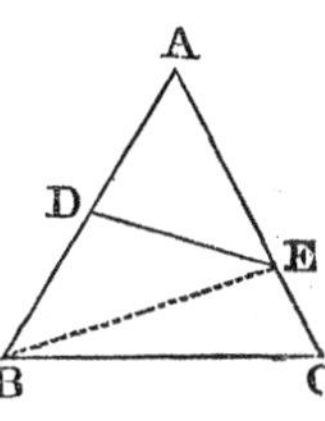

Let the two triangles ABC, ADE have the angle A in common; then will the triangle ABC be to the triangle ADE as the rectangle AB × AC is to the rectangle AD × AE.

Join BE. Then the two triangles ABE, ADE, having the common vertex E, have the same altitude, and are to each other as their bases AB, AD (Prop. VI., Cor. 1); therefore

$$ABE : ADE :: AB : AD.$$

Also, the two triangles ABC, ABE, having the common vertex B, have the same altitude, and are to each other as their bases AC, AE; therefore

$$ABC : ABE :: AC : AE.$$

Hence (Prop. XI., Cor., B. II.).

$$ABC : ADE :: AB \times AC : AD \times AE.$$

Therefore, two triangles, &c.

Cor. 1. If the rectangles of the sides containing the equal angles are equivalent, the triangles will be equivalent.

Cor. 2. Equiangular parallelograms are to each other as the rectangles of the sides which contain the equal angles.

PROPOSITION XXIV. THEOREM.

Similar triangles are to each other as the squares described on their homologous sides.

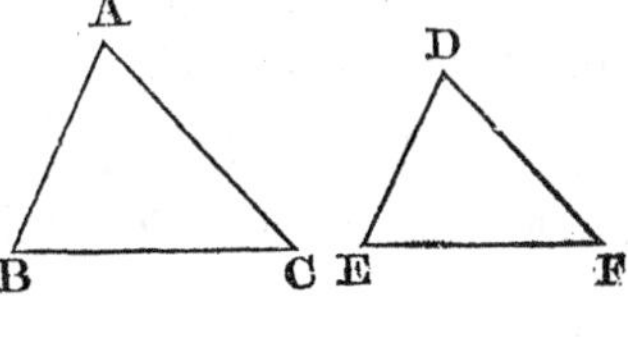

Let ABC, DEF be two similar triangles, having the angle A equal to D, the angle B equal to E, and C equal to F; then the triangle ABC is to the triangle DEF as the square on BC is to the square on EF.

By similar triangles, we have (Def. 3)

$$AB : DE :: BC : EF.$$

Also, $$BC : EF :: BC : EF.$$

Multiplying together the corresponding terms of these proportions, we obtain (Prop. XI., B. II.),

$$AB \times BC : DE \times EF :: BC^2 : EF^2.$$

But, by Prop. XXIII.,

$$ABC : DEF :: AB \times BC : DE \times EF;$$

hence (Prop. IV., B. II.)

$$ABC : DEF :: BC^2 : EF^2.$$

Therefore, similar triangles, &c.

PROPOSITION XXV. THEOREM.

Two similar polygons may be divided into the same number of triangles, similar each to each, and similarly situated.

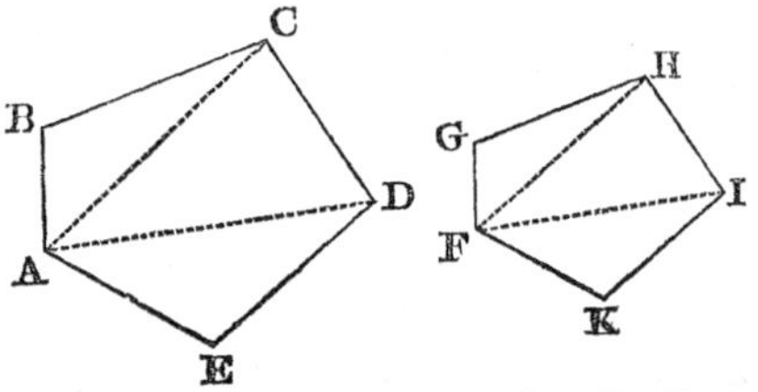

Let ABCDE, FGHIK be two similar polygons; they may be divided into the same number of similar triangles. Join AC, AD, FH, FI.

Because the polygon ABCDE is similar to the polygon FGHIK, the angle B is equal to the angle G (Def. 3), and AB : BC : : FG : GH. And, because the triangles ABC, FGH have an angle in the one equal to an angle in the other, and the sides about these equal angles proportional, they are similar (Prop. XX.); therefore the angle BCA is equal to the angle GHF. Also, because the polygons are similar, the whole angle BCD is equal (Def. 3) to the whole angle GHI; therefore, the remaining angle ACD is equal to the remaining angle FHI. Now, because the triangles ABC FGH are similar,

$$AC : FH :: BC : GH.$$

And, because the polygons are similar (Def. 3),

$$BC : GH :: CD : HI;$$

whence

$$AC : FH :: CD : HI;$$

that is, the sides about the equal angles ACD, FHI are proportional; therefore the triangle ACD is similar to the triangle FHI (Prop. XX.). For the same reason, the triangle ADE is similar to the triangle FIK; therefore the similar polygons ABCDE, FGHIK are divided into the same number of triangles, which are similar, each to each, and similarly situated.

Cor. Conversely, *if two polygons are composed of the same number of triangles, similar and similarly situated, the polygons are similar.*

For, because the triangles are similar, the angle ABC is equal to FGH; and because the angle BCA is equal to GHF, and ACD to FHI, therefore the angle BCD is equal to GHI. For the same reason, the angle CDE is equal to HIK, and so on for the other angles. Therefore the two polygons are mutually equiangular.

Moreover, the sides about the equal angles are proportional. For, because the triangles are similar, AB : FG : : BC : GH. Also, BC : GH : : AC : FH, and AC : FH : : CD : HI; hence BC : GH : : CD : HI. In the same manner, it may be proved that CD : HI : : DE : IK, and so on for the other sides. Therefore the two polygons are similar.

PROPOSITION XXVI. THEOREM.

The perimeters of similar polygons are to each other as their homologous sides; and their areas are as the squares of those sides.

Let ABCDE, FGHIK be two similar polygons, and let AB be the side homologous to FG; then the perimeter of ABCDE is to the perimeter of FGHIK as AB is to FG; and the area of ABCDE is to the area of FGHIK as AB^2 is to FG^2

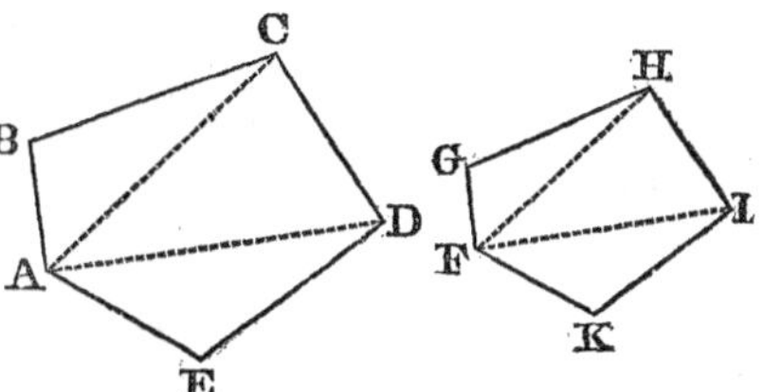

First. Because the polygon ABCDE is similar to the polygon FGHIK (Def. 3),

AB : FG : : BC : GH : : CD : HI, &c.;

therefore (Prop. IX., B. II.) the sum of the antecedents AB +BC+CD, &c., which form the perimeter of the first figure, is to the sum of the consequents FG+GH+HI, &c., which form the perimeter of the second figure, as any one antecedent is to its consequent, or as AB to FG.

Secondly. Because the triangle ABC is similar to the triangle FGH, the triangle ABC : triangle FGH : : AC^2 : FH^2 (Prop. XXIV.).

And, because the triangle ACD is similar to the triangle FHI,

ACD : FHI : : AC^2 : FH^2.

Therefore the triangle ABC : triangle FGH : : triangle ACD : triangle FHI (Prop. IV., B. II.). In the same manner, it may be proved that

ACD : FHI : : ADE : FIK.

Therefore, as the sum of the antecedents ABC+ACD+ADE, or the polygon ABCDE, is to the sum of the consequents FGH+FHI+FIK, or the polygon FGHIK, so is any one antecedent, as ABC, to its consequent FGH; or, as AB^2 to FG^2. Therefore, similar polygons, &c.

PROPOSITION XXVII. THEOREM.

If two chords in a circle intersect each other, the rectangle contained by the parts of the one, is equal to the rectangle contained by the parts of the other.

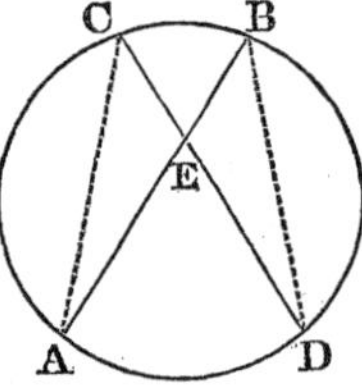

Let the two chords AB, CD in the circle ACBD, intersect each other in the point E; the rectangle contained by AE, EB is equal to the rectangle contained by DE, EC.

Join AC and BD. Then, in the triangles ACE, DBE, the angles at E are equal, being vertical angles (Prop. V., B. I.); the angle A is equal to the angle D, being inscribed in the same segment (Prop. XV., Cor. 1., B. III.); therefore the angle C is equal to the angle B. The triangles are consequently similar; and hence (Prop. XVIII.)

$$AE : DE :: EC : EB,$$

or (Prop. I., B. II.),

$$AE \times EB = DE \times EC.$$

Therefore, if two chords, &c.

Cor. The parts of two chords which intersect each other in a circle are reciprocally proportional; that is, AE : DE :: EC : EB.

PROPOSITION XXVIII. THEOREM.

If from a point without a circle, a tangent and a secant be drawn, the square of the tangent will be equivalent to the rectangle contained by the whole secant and its external segment.

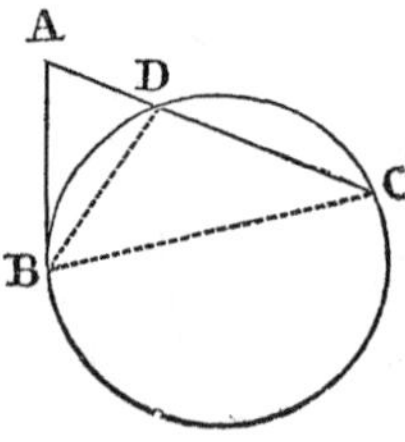

Let A be any point without the circle BCD, and let AB be a tangent, and AC a secant; then the square of AB is equivalent to the rectangle AD×AC.

Join BD and BC. Then the triangles ABD and ABC are similar; because they have the angle A in common; also, the angle ABD formed by a tangent and a chord is measured by half the arc BD

(Prop. XVI., B. III.); and the angle C is measured by half the same arc, therefore the angle ABD is equal to C, and the two triangles ABD, ABC are equiangular, and, consequently, similar; therefore (Prop. XVIII.)

$$AC : AB :: AB : AD;$$

whence (Prop. I., B. II.),

$$AB^2 = AC \times AD.$$

Therefore, if from a point, &c.

Cor. 1. If from a point without a circle, a tangent and a secant be drawn, the tangent will be a mean proportional between the secant and its external segment.

Cor. 2. If from a point without a circle, two secants be drawn, the rectangles contained by the whole secants and their external segments will be equivalent to each other; for each of these rectangles is equivalent to the square of the tangent from the same point.

Cor. 3. If from a point without a circle, two secants be drawn, the whole secants will be reciprocally proportional to their external segments.

PROPOSITION XXIX. THEOREM.

If an angle of a triangle be bisected by a line which cuts the base, the rectangle contained by the sides of the triangle, is equivalent to the rectangle contained by the segments of the base, together with the square of the bisecting line.

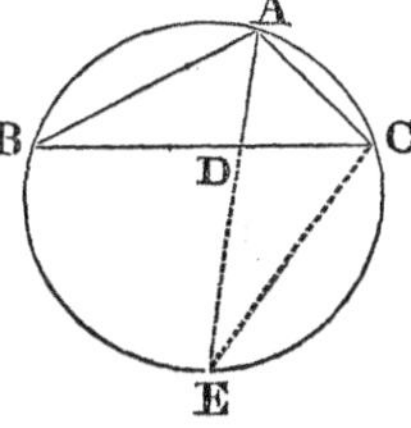

Let ABC be a triangle, and let the angle BAC be bisected by the straight line AD; the rectangle BA×AC is equivalent to BD×DC together with the square of AD.

Describe the circle ACEB about the triangle, and produce AD to meet the circumference in E, and join EC. Then, because the angle BAD is equal to the angle CAE, and the angle ABD to the angle AEC, for they are in the same segment (Prop. XV., Cor. 1, B. III.), the triangles ABD, AEC are mutually equiangular and similar; therefore (Prop. XVIII.)

$$BA : AD :: EA : AC;$$

consequently (Prop. I., B. II.),

$$BA \times AC = AD \times AE.$$

But AE=AD+DE; and multiplying each of these equals by AD, we have (Prop. III.) $AD \times AE = AD^2 + AD \times DE$. But $AD \times DE = BD \times DC$ (Prop. XXVII.); hence

$$BA \times AC = BD \times DC + AD^2.$$

Therefore, if an angle, &c.

PROPOSITION XXX. THEOREM.

The rectangle contained by the diagonals of a quadrilateral inscribed in a circle, is equivalent to the sum of the rectangles of the opposite sides.

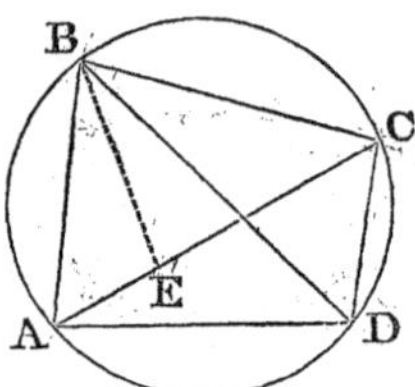

Let ABCD be any quadrilateral inscribed in a circle, and let the diagonals AC, BD be drawn; the rectangle AC×BD is equivalent to the sum of the two rectangles AD×BC and AB×CD.

Draw the straight line BE, making the angle ABE equal to the angle DBC. To each of these equals add the angle EBD; then will the angle ABD be equal to the angle EBC. But the angle BDA is equal to the angle BCE, because they are both in the same segment (Prop. XV., Cor. 1, B. III.); hence the triangle ABD is equiangular and similar to the triangle EBC. Therefore we have

AD : BD : : CE : BC;

and, consequently, AD×BC=BD×CE.

Again, because the angle ABE is equal to the angle DBC and the angle BAE to the angle BDC, being angles in the same segment, the triangle ABE is similar to the triangle DBC; and hence

AB : AE : : BD : CD;

consequently, AB×CD=BD×AE.

Adding together these two results, we obtain

AD×BC+AB×CD=BD×CE+BD×AE,

which equals BD×(CE+AE), or BD×AC.

Therefore, the rectangle, &c.

PROPOSITION XXXI. THEOREM.

If from any angle of a triangle, a perpendicular be drawn to the opposite side or base, the rectangle contained by the sum and difference of the other two sides, is equivalent to the rectangle contained by the sum and difference of the segments of the base

Let ABC be any triangle, and let AD be a perpendicular drawn from the angle A on the base BC; then

(AC+AB)×(AC−AB)=(CD+DB)×(CD−DB).

From A as a center, with a radius equal to AB, the short-

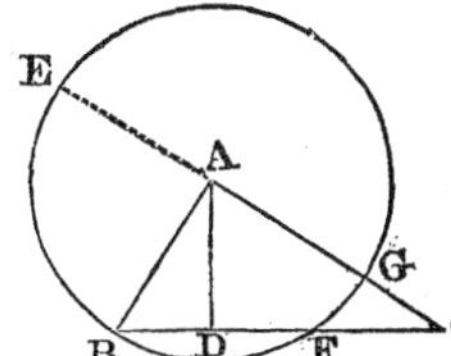

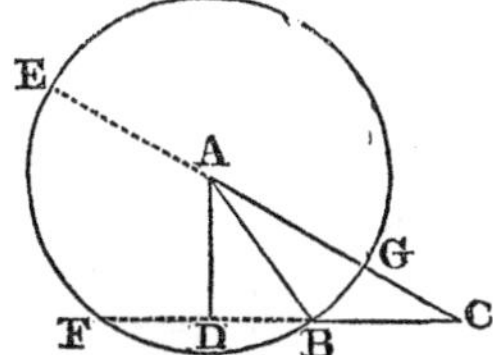

er of the two sides, describe a circumference BFE. Produce AC to meet the circumference in E, and CB, if necessary, to meet it in F.

Then, because AB is equal to AE or AG, CE=AC+AB, the sum of the sides; and CG=AC−AB, the difference of the sides. Also, because BD is equal to DF (Prop. VI., B. III.); when the perpendicular falls within the triangle, CF=CD−DF=CD−DB, the difference of the segments of the base. But when the perpendicular falls without the triangle, CF=CD+DF=CD+DB, the sum of the segments of the base.

Now in either case, the rectangle CE×CG is equivalent to CB×CF (Prop. XXVIII., Cor. 2); that is,

$$(AC+AB)\times(AC-AB)=(CD+DB)\times(CD-DB).$$

Therefore, if from any angle, &c.

Cor. If we reduce the preceding equation to a proportion (Prop. II., B. II.), we shall have

$$BC : AC+AB :: AC-AB : CD-DB;$$

that is, the base of any triangle is to the sum of the two other sides, as the difference of the latter is to the difference of the segments of the base made by the perpendicular.

PROPOSITION XXXII. THEOREM.

The diagonal and side of a square have no common measure.

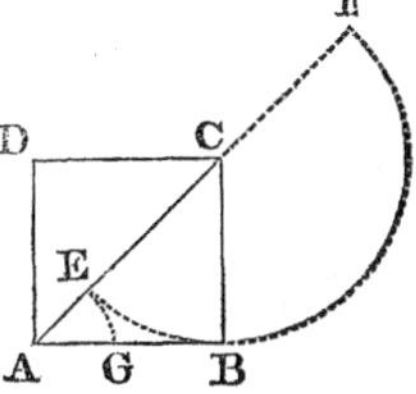

Let ABCD be a square, and AC its diagonal; AC and AB have no common measure.

In order to find the common measure, if there is one, we must apply CB to CA as often as it is contained in it. For this purpose, from the center C, with a radius CB, describe the semicircle EBF. We perceive that CB is contained once in AC, with a remainder AE, which remainder must be compared with BC or its equal AB.

Now, since the angle ABC is a right angle, AB is a tangent to the circumference; and AE : AB :: AB : AF (Prop.

XXVIII., Cor. 1). Instead, therefore, of comparing AE with AB, we may substitute the equal ratio of AB to AF. But AB is contained twice in AF, with a remainder AE, which must be again compared with AB. Instead, however, of comparing AE with AB, we may again employ the equal ratio of AB to AF. Hence at each operation we are obliged to compare AB with AF, which leaves a remainder AE; from which we see that the process will never terminate, and therefore there is no common measure between the diagonal and side of a square that is, there is no line which is contained an exact number of times in each of them.

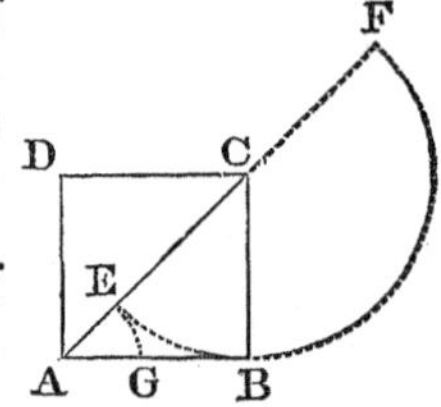

BOOK V

PROBLEMS

Postulates.

1. A straight line may be drawn from any one point to any other point.
2. A terminated straight line may be produced to any length in a straight line.
3. From the greater of two straight lines, a part may be cut off equal to the less.
4. A circumference may be described from any center, and with any radius.

PROBLEM I.

To bisect a given straight line.

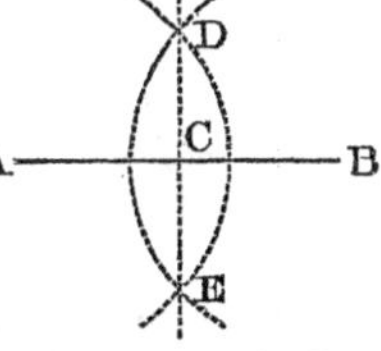

Let AB be the given straight line which it is required to bisect.

From the center A, with a radius greater than the half of AB, describe an arc of a circle (Postulate 4); and from the center B, with the same radius, describe another arc intersecting the former in D and E. Through the points of intersection, draw the straight line DE (Post. 1); it will bisect AB in C.

For, the two points D and E, being each equally distant from the extremities A and B, must both lie in the perpendicular, raised from the middle point of AB (Prop. XVIII. Cor., B. I.). Therefore the line DE divides the line AB into two equal parts at the point C.

PROBLEM II.

To draw a perpendicular to a straight line, from a given point in that line.

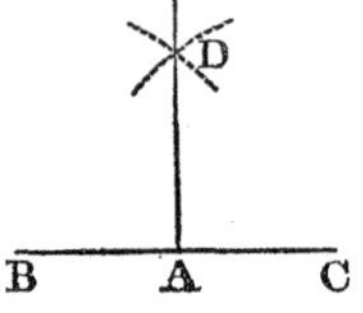

Let BC be the given straight line, and A the point given in it; it is required to draw a straight line perpendicular to BC through the given point A.

In the straight line BC take any point B, and make AC equal to AB (Post. 3). From B as a center, with a radius greater than BA, describe an arc of a circle (Post. 4); and from C as a center, with the same radius, describe another arc intersecting the former in D. Draw AD (Post. 1), and it will be the perpendicular required.

For, the points A and D, being equally distant from B and C, must be in a line perpendicular to the middle of BC (Prop. XVIII., Cor., B. I.). Therefore AD has been drawn perpendicular to BC from the point A.

Scholium. The same construction serves to make a right angle BAD at a given point A, on a given line BC.

PROBLEM III.

To draw a perpendicular to a straight line, from a given point without it.

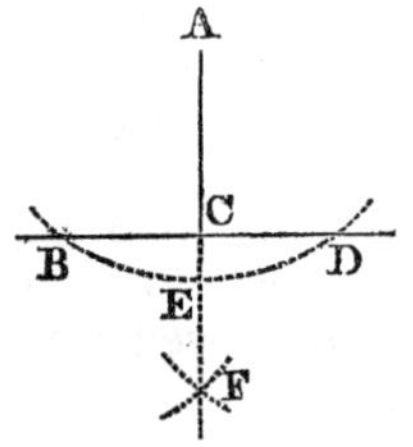

Let BD be a straight line of unlimited length, and let A be a given point without it. It is required to draw a perpendicular to BD from the point A.

Take any point E upon the other side of BD; and from the center A, with the radius AE, describe the arc BD cutting the line BCD in the two points B and D. From the points B and D as centers, describe two arcs, as in Prob. II., cutting each other in F. Join AF, and it will be the perpendicular required.

For the two points A and F are each equally distant from the points B and D; therefore the line AF has been drawn perpendicular to BD (Prop. XVIII., Cor., B. I.), from the given point A.

PROBLEM IV.

At a given point in a straight line, to make an angle equal to a given angle.

Let AB be the given straight line, A the given point in it, and C the given angle; it is required to make an angle at the point A in the straight line AB, that shall be equal to the given angle C.

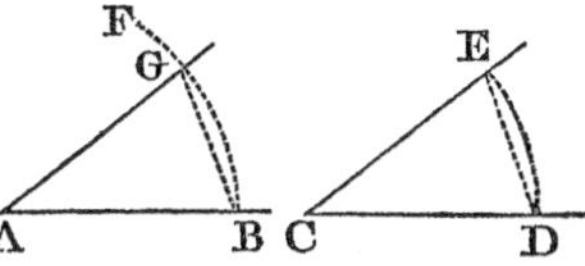

With C as a center, and any radius, describe an arc DE terminating in the sides of the angle; and from the point A as a center, with the same radius, describe the indefinite arc BF. Draw the chord DE; and from B as a center, with a radius equal to DE, describe an arc cutting the arc BF in G. Draw AG, and the angle BAG will be equal to the given angle C.

For the two arcs BG, DE are described with equal radii, and they have equal chords; they are, therefore, equal (Prop. III., B. III.). But equal arcs subtend equal angles (Prop IV., B. III.); and hence the angle A has been made equal to the given angle C.

PROBLEM V.

To bisect a given arc or angle.

First. Let ADB be the given arc which it is required to bisect.

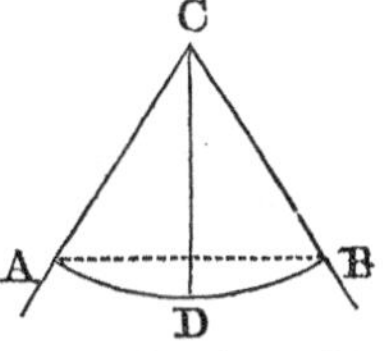

Draw the chord AB, and from the center C draw CD perpendicular to AB (Prob. III.); it will bisect the arc ADB (Prop. VI., B. III.), because CD is a radius perpendicular to a chord.

Secondly. Let ACB be an angle which it is required to bisect. From C as a center, with any radius, describe an arc AB; and, by the first case, draw the line CD bisecting the arc ADB. The line CD will also bisect the angle ACB. For the angles ACD, BCD are equal, being subtended by the equal arcs AD, DB (Prop. IV., B. III.).

Scholium. By the same construction, each of the halves AD, DB may be bisected; and thus by successive bisections an arc or angle may be divided into four equal parts, into eight, sixteen, &c.

PROBLEM VI.

Through a given point, to draw a straight line parallel to a given line.

Let A be the given point, and BC the given straight line; it is required to draw through the point A, a straight line parallel to BC.

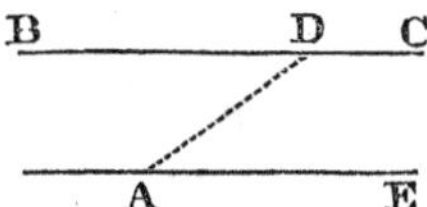

In BC take any point D, and join AD. Then at the point A, in the straight line AD, make the angle DAE equal to the angle ADB (Prob. IV.).

Now, because the straight line AD, which meets the two straight lines BC, AE, makes the alternate angles ADB, DAE equal to each other, AE is parallel to BC (Prop. XXII., B. I.). Therefore the straight line AE has been drawn through the point A, parallel to the given line BC.

PROBLEM VII.

Two angles of a triangle being given, to find the third angle.

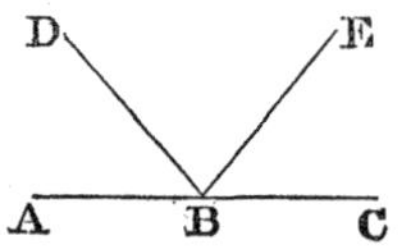

The three angles of every triangle are together equal to two right angles (Prop. XXVII., B. I.). Therefore, draw the indefinite line ABC. At the point B make the angle ABD equal to one of the given angles (Prob. IV.), and the angle DBE equal to the other given angle; then will the angle EBC be equal to the third angle of the triangle. For the three angles ABD, DBE, EBC are together equal to two right angles (Prop. II., B I.), which is the sum of all the angles of the triangle.

PROBLEM VIII.

Given two sides and the included angle of a triangle, to construct the triangle.

Draw the straight line BC equal to one of the given sides. At the point B make the angle ABC equal to the given angle (Prob. IV.); and take AB equal to the other given side. Join AC, and ABC will be the

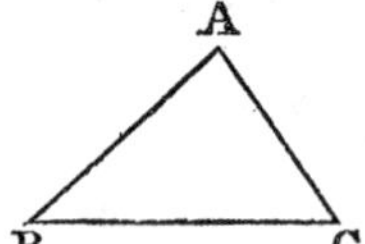

triangle required. For its sides AB, BC are made equal to the given sides, and the included angle B is made equal to the given angle.

PROBLEM IX.

Given one side and two angles of a triangle, to construct the triangle.

The two given angles will either be both adjacent to the given side, or one adjacent and the other opposite. In the latter case, find the third angle (Prob. VII.); and then the two adjacent angles will be known.

Draw the straight line AB equal to the given side; at the point A make the angle BAC equal to one of the adjacent angles; and at the point B make the angle ABD equal to the other adjacent angle. The two lines AC, BD will cut each other in E, and ABE will be the triangle required; for its side AB is equal to the given side, and two of its angles are equal to the given angles.

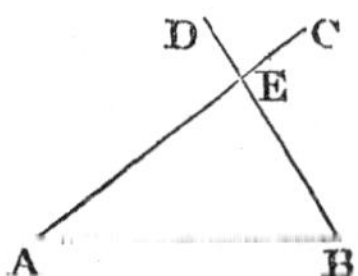

PROBLEM X.

Given the three sides of a triangle, to construct the triangle

Draw the straight line BC equal to one of the given sides. From the point B as a center, with a radius equal to one of the other sides, describe an arc of a circle; and from the point C as a center, with a radius equal to the third side, describe another arc cutting the former in A. Draw AB, AC; then will ABC be the triangle required, because its three sides are equal to the three given straight lines.

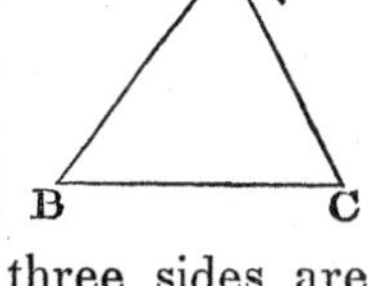

Scholium. If one of the given lines was greater than the sum of the other two, the arcs would not intersect each other, and the problem would be impossible; but the solution will always be possible when the sum of any two sides is greater than the third.

PROBLEM XI.

Given two sides of a triangle, and an angle opposite one of them, to construct the triangle.

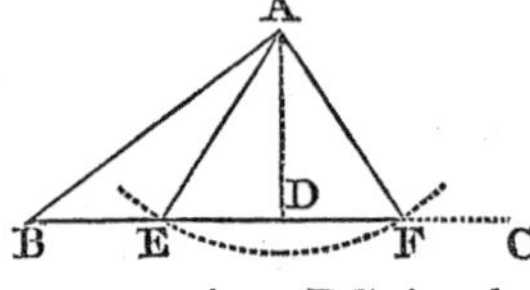

Draw an indefinite straight line BC. At the point B make the angle ABC equal to the given angle, and make BA equal to that side which is adjacent to the given angle. Then from A as a center, with a radius equal to the other side, describe an arc cutting BC in the points E and F. Join AE, AF. If the points E and F both fall on the same side of the angle B, each of the triangles ABE, ABF will satisfy the given conditions; but if they fall upon different sides of B, only one of them, as ABF, will satisfy the conditions, and therefore this will be the triangle required.

If the points E and F coincide with one another, which will happen when AEB is a right angle, there will be only one triangle ABD, which is the triangle required.

Scholium. If the side opposite the given angle were less than the perpendicular let fall from A upon BC, the problem would be impossible.

PROBLEM XII.

Given two adjacent sides of a parallelogram, and the included angle, to construct the parallelogram.

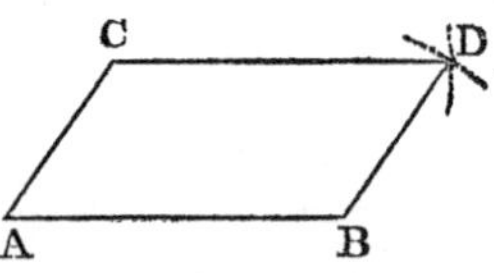

Draw the straight line AB equal to one of the given sides. At the point A make the angle BAC equal to the given angle; and take AC equal to the other given side. From the point C as a center, with a radius equal to AB, describe an arc; and from the point B as a center, with a radius equal to AC, describe another arc intersecting the former in D. Draw BD, CD; then will ABDC be the parallelogram required.

For, by construction, the opposite sides are equal; therefore the figure is a parallelogram (Prop. XXX., B. I.), and it is formed with the given sides and the given angle

Cor. If the given angle is a right angle, the figure will be a rectangle; and if, at the same time, the sides are equal, it will be a square.

PROBLEM XIII.

To find the center of a given circle or arc.

Let ABC be the given circle or arc; it is required to find its center.

Take any three points in the arc, as A B, C, and join AB, BC. Bisect AB in D (Prob. I.), and through D draw DF perpendicular to AB (Prob. II.). In the same manner, draw EF perpendicular to BC at its middle point. The perpendiculars DF, EF will meet in a point F equally distant from the points A, B, and C (Prop. VII., B. III.); and therefore F is the center of the circle.

Scholium. By the same construction, a circumference may be made to pass through three given points A, B, C; and also, a circle may be described about a triangle.

PROBLEM XIV.

Through a given point, to draw a tangent to a given circle

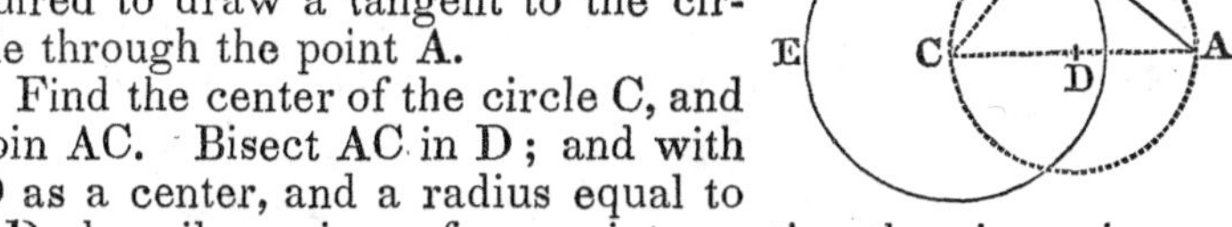

First. Let the given point A be without the circle BDE; it is required to draw a tangent to the circle through the point A.

Find the center of the circle C, and join AC. Bisect AC in D; and with D as a center, and a radius equal to AD, describe a circumference intersecting the given circumference in B. Draw AB, and it will be the tangent required.

Draw the radius CB. The angle ABC, being inscribed in a semicircle is a right angle (Prop. XV., Cor. 2, B. III.). Hence the line AB is a perpendicular at the extremity of the radius CB; it is, therefore, a tangent to the circumference (Prop IX., B. III.).

Secondly. If the given point is in the circumference of the circle, as the point B, draw the radius BC, and make BA perpendicular to BC. BA will be the tangent required (Prop. IX., B. III.).

Scholium. When the point A lies without the circle, two tangents may always be drawn; for the circumference whose center is D intersects the given circumference in two points.

PROBLEM XV.

To inscribe a circle in a given triangle.

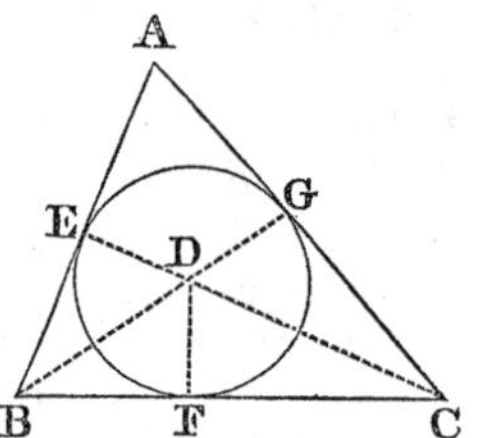

Let ABC be the given triangle; it is required to inscribe a circle in it.

Bisect the angles B and C by the lines BD, CD, meeting each other in the point D. From the point of intersection, let fall the perpendiculars DE, DF, DG on the three sides of the triangle; these perpendiculars will all be equal. For, by construction, the angle EBD is equal to the angle FBD; the right angle DEB is equal to the right angle DFB; hence the third angle BDE is equal to the third angle BDF (Prop. XXVII., Cor. 2, B. I.). Moreover, the side BD is common to the two triangles BDE, BDF, and the angles adjacent to the common side are equal; therefore the two triangles are equal, and DE is equal to DF. For the same reason, DG is equal to DF. Therefore the three straight lines DE, DF, DG are equal to each other; and if a circumference be described from the center D, with a radius equal to DE, it will pass through the extremities of the lines DF, DG. It will also touch the straight lines AB, BC, CA, because the angles at the points E, F, G are right angles (Prop. IX., B. III.). Therefore the circle EFG is inscribed in the triangle ABC (Def. 11, B. III.)

Scholium. The three lines which bisect the angles of a triangle, all meet in the same point, viz., the center of the in scribed circle.

PROBLEM XVI.

Upon a given straight line, to describe a segment of a circle which shall contain a given angle.

Let AB be the given straight line, upon which it is required to describe a segment of a circle containing a given angle.

At the point A, in the straight line AB, make the angle BAD equal to the given angle; and from the point A draw

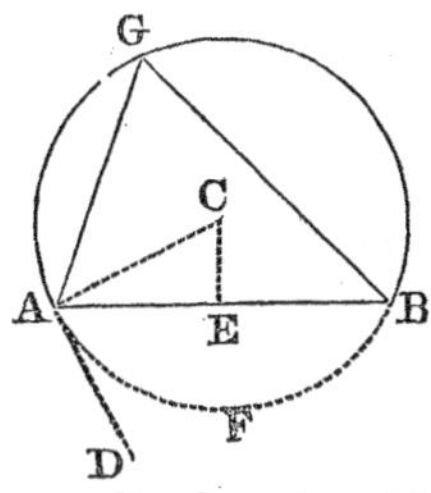

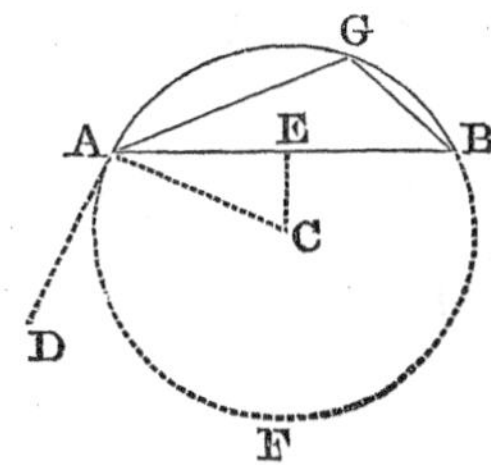

AC perpendicular to AD. Bisect AB in E, and from E draw EC perpendicular to AB. From the point C, where these perpendiculars meet, with a radius equal to AC, de scribe a circle. Then will AGB be the segment required.

For, since AD is a perpendicular at the extremity of the radius AC, it is a tangent (Prop. IX., B. III.); and the angle BAD is measured by half the arc AFB (Prop. XVI., B. III.). Also, the angle AGB, being an inscribed angle, is measured by half the same arc AFB; hence the angle AGB is equal to the angle BAD, which, by construction, is equal to the given angle. Therefore all the angles inscribed in the segment AGB are equal to the given angle.

Scholium. If the given angle was a right angle, the required segment would be a semicircle, described on AB as a diameter.

PROBLEM XVII.

To divide a given straight line into any number of equal parts, or into parts proportional to given lines.

First. Let AB be the given straight line which it is proposed to divide into any number of equal parts, as, for example, five.

From the point A draw the indefinite straight line AC, making any angle with AB. In AC take any point D, and set off AD five times upon AC. Join BC, and draw DE parallel to it; then is AE the fifth part of AB.

For, since ED is parallel to BC, AE : AB : : AD : AC (Prop. XVI., B. IV.). But AD is the fifth part of AC; therefore AE is the fifth part of AB.

Secondly. Let AB be the given straight line, and AC a divided line; it is required to divide AB similarly to AC. Suppose AC to be divided in the points D and E. Place AB, AC so as to contain any angle; join BC, and through the points D, E draw DF, EG parallel to BC. The line AB wil. be divided into parts proportional to those of AC.

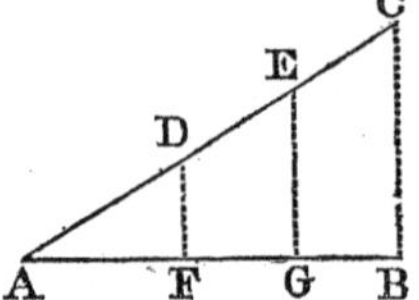

For, because DF and EG are both parallel to CB, we have AD : AF : : DE : FG : EC : GB (Prop. XVI., Cor. 2, B. IV.).

PROBLEM XVIII.

To find a fourth proportional to three given lines.

From any point A draw two straight lines AD, AE, containing any angle DAE; and make AB, BD, AC respectively equal to the proposed lines. Join B, C; and through D draw DE parallel to BC; then will CE be the fourth proportional required.

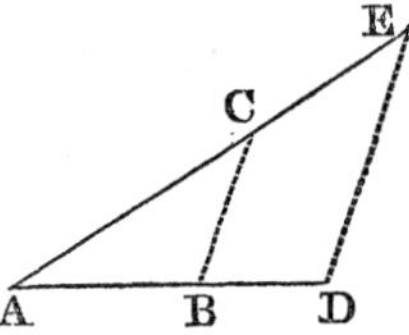

For, because BC is parallel to DE, we have

AB : BD : : AC : CE (Prop. XVI., B. IV.).

Cor. In the same manner may be found a third proportional to two given lines A and B; for this will be the same as a fourth proportional to the three lines A, B, B.

PROBLEM XIX.

To find a mean proportional between two given lines.

Let AB, BC be the two given straight lines; it is required to find a mean proportional between them.

Place AB, BC in a straight line; upon AC describe the semicircle ADC; and from the point B draw BD perpendicular to AC. Then will BD be the mean proportional required.

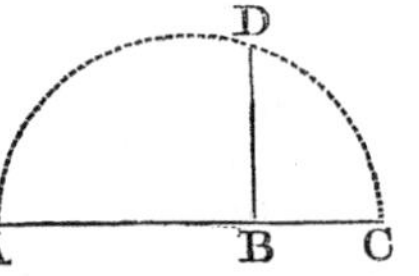

For the perpendicular BD, let fall from a point in the circumference upon the diameter, is a mean proportional between the two segments of the diameter AB, BC (Prop. XXII., Cor., B. IV.); and these segments are equal to the two given lines.

PROBLEM XX.

To divide a given line into two parts, such that the greater part may be a mean proportional between the whole line and the other part.

Let AB be the given straight line; it is required to divide it into two parts at the point F, such that AB : AF : : AF : FB.

At the extremity of the line AB, erect the perpendicular BC, and make it equal to the half of AB. From C as a center, with a radius equal to CB, describe a circle. Draw AC cutting the circumference in D; and make AF equal to AD. The line AB will be divided in the point F in the manner required.

For, since AB is a perpendicular to the radius CB at its extremity, it is a tangent (Prop. IX., B. III.); and if we produce AC to E, we shall have AE : AB : : AB : AD (Prop. XXVIII., B. IV.). Therefore, by division (Prop. VII., B. II.), AE—AB : AB : : AB—AD : AD. But, by construction, AB is equal to DE; and therefore AE—AB is equal to AD or AF; and AB—AD is equal to FB. Hence AF : AB : : FB : AD or AF; and, consequently, by inversion (Prop. V B. II.),

AB : AF : : AF : FB.

Scholium. The line AB is said to be divided in *extreme and mean ratio.* An example of its use may be seen in Prop. V., Book VI.

PROBLEM XXI.

Through a given point in a given angle, to draw a straight line so that the parts included between the point and the sides of the angle, may be equal.

Let A be the given point, and BCD the given angle; it is required to draw through A a line BD, so that BA may be equal to AD.

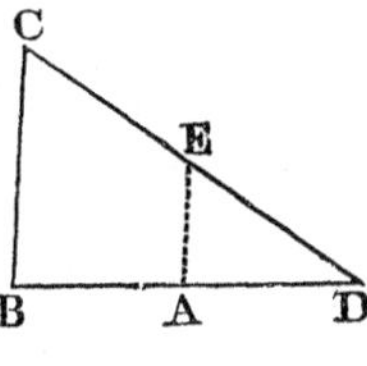

Through the point A draw AE parallel to BC; and take DE equal to CE. Through the points D and A draw the line BAD; it will be the line required.

For, because AE is parallel to BC we have (Prop. XVI B. IV.),

DE : EC : : DA : AB.

But DE is equal to EC; therefore DA is equal to AB.

PROBLEM XXII.

To describe a square that shall be equivalent to a given parallelogram, or to a given triangle.

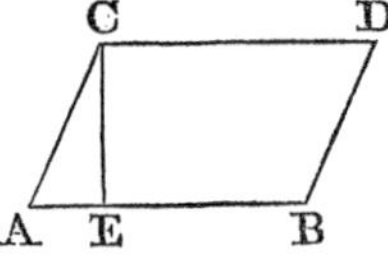

First. Let ABDC be the given parallelogram, AB its base, and CE its altitude. Find a mean proportional between AB and CE (Prob. XIX.), and represent it by X; the square described on X will be equivalent to the given parallelogram ABDC.

For, by construction, AB : X : : X : CE; hence X^2 is equal to AB × CE (Prop. I., Cor., B. II.). But AB × CE is the measure of the parallelogram; and X^2 is the measure of the square. Therefore the square described on X is equivalent to the given parallelogram ABDC.

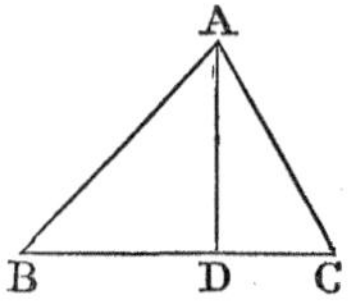

Secondly. Let ABC be the given triangle, BC its base, and AD its altitude. Find a mean proportional between BC and the half of AD, and represent it by Y. Then will the square described on Y be equivalent to the triangle ABC.

For, by construction, BC : Y : : Y : $\frac{1}{2}$ AD; hence Y^2 is equivalent to BC × $\frac{1}{2}$ AD. But BC × $\frac{1}{2}$ AD is the measure of the triangle ABC; therefore the square described on Y is equivalent to the triangle ABC.

PROBLEM XXIII.

Upon a given line, to construct a rectangle equivalent to a given rectangle.

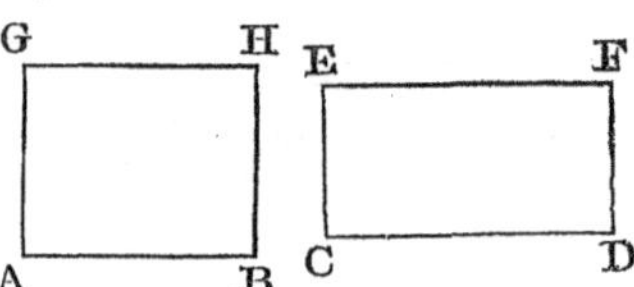

Let AB be the given straight line, and CDFE the given rectangle. It is required to construct on the line AB a rectangle equivalent to CDFE.

Find a fourth proportional (Prob. XVIII.) to the three lines AB, CD, CE, and let AG be that fourth proportional. The rectangle constructed on the lines AB, AG will be equivalent to CDFE.

For, because AB : CD : : CE : AG, by Prop. I., B. II., AB×AG=CD×CE. Therefore the rectangle ABHG is equivalent to the rectangle CDFE; and it is constructed upon the given line AB.

PROBLEM XXIV.

To construct a triangle which shall be equivalent to a given polygon.

Let ABCDE be the given polygon; it is required to construct a triangle equivalent to it.

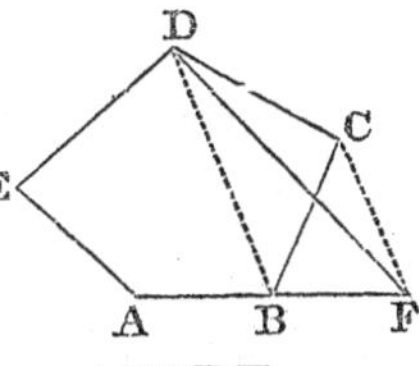

Draw the diagonal BD cutting off the triangle BCD. Through the point C, draw CF parallel to DB, meeting AB produced in F. Join DF; and the polygon AFDE will be equivalent to the polygon ABCDE.

For the triangles BFD, BCD, being upon the same base BD, and between the same parallels BD, FC, are equivalent. To each of these equals, add the polygon ABDE; then will the polygon AFDE be equivalent to the polygon ABCDE; that is, we have found a polygon equivalent to the given polygon, and having the number of its sides diminished by one.

In the same manner, a polygon may be found equivalent to AFDE, and having the number of its sides diminished by one; and, by continuing the process, the number of sides may be at last reduced to three, and a triangle be thus obtained equivalent to the given polygon.

PROBLEM XXV.

To make a square equivalent to the sum or difference of two given squares.

First. To make a square equivalent to the sum of two given squares. Draw two indefinite lines AB, BC at right angles to each other. Take AB equal to the side of one of the given squares, and BC equal to the side of the other. Join AC; it will be the side of the required square.

For the triangle ABC, being right-angled at B, the square

on AC will be equivalent to the sum of the squares upon AB and BC (Prop. XI., B. IV.).

Secondly. To make a square equivalent to the difference of two given squares. Draw the lines AB, BC at right angles to each other; and take AB equal to the side of the less square. Then from A as a center, with a radius equal to the side of the other square, describe an arc intersecting BC in C; BC will be the side of the square required; because the square of BC is equivalent to the difference of the squares of AC and AB (Prop. XI., Cor. 1, B. IV.).

Scholium. In the same manner, a square may be made equivalent to the sum of three or more given squares; for the same construction which reduces two of them to one will reduce three of them to two, and these two to one.

PROBLEM XXVI.

Upon a given straight line, to construct a polygon simila to a given polygon.

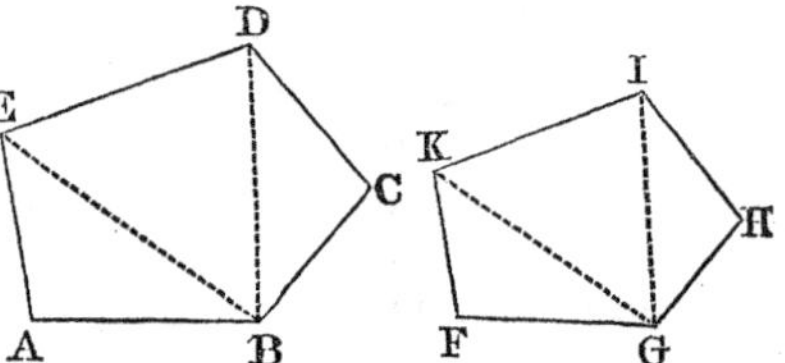

Let ABCDE be the given polygon, and FG be the given straight line; it is required upon the line FG to construct a polygon similar to ABCDE.

Draw the diagonals BD, BE. At the point F, in the straight line FG, make the angle GFK equal to the angle BAE; and at the point G make the angle FGK equal to the angle ABE. The lines FK, GK will intersect in K, and FGK will be a triangle similar to ABE. In the same manner, on GK construct the triangle GKI similar to BED, and on GI construct the triangle GIH similar to BDC. The polygon FGHIK will be the polygon required. For these two polygons are composed of the same number of triangles, which are similar to each other, and similarly situated; therefore the polygons are similar (Prop. XXV., Cor., B. IV.)

PROBLEM XXVII.

Given the area of a rectangle, and the sum of two adjacent sides, to construct the rectangle.

Let AB be a straight line equal to the sum of the sides of the required rectangle.

Upon AB as a diameter, describe a semicircle. At the point A erect the perpendicular AC, and make it equal to the side of a square having the given area. Through C draw the line CD parallel to AB, and let it meet the circumference in D; and from D draw DE perpendicular to AB. Then will AE and EB be the sides of the rectangle required.

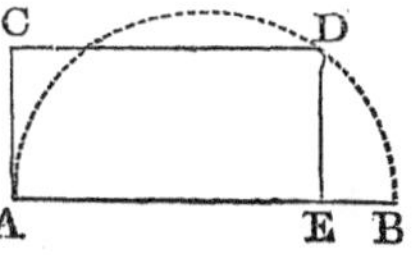

For, by Prop. XXII., Cor., B. IV., the rectangle AE×EB is equivalent to the square of DE or CA, which is, by construction, equivalent to the given area. Also, the sum of the sides AE and EB is equal to the given line AB.

Scholium. The side of the square having the given area, must not be greater than the half of AB; for in that case the line CD would not meet the circumference ADB.

PROBLEM XXVIII.

Given the area of a rectangle, and the difference of two adjacent sides, to construct the rectangle.

Let AB be a straight line equal to the difference of the sides of the required rectangle.

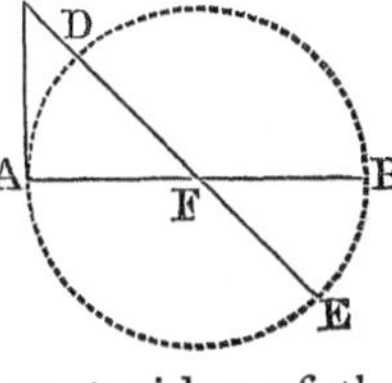

Upon AB as a diameter, describe a circle; and at the extremity of the diameter, draw the tangent AC equal to the side of a square having the given area. Through the point C and the center F draw the secant CE; then will CD, CE be the adjacent sides of the rectangle required.

For, by Prop. XXVIII., B. IV., the rectangle CD×CE is equivalent to the square of AC, which is, by construction, equivalent to the given area. Also, the difference of the lines CE, CD is equal to DE or AB.

BOOK VI.

REGULAR POLYGONS, AND THE AREA OF THE CIRCLE.

Definition.

A *regular polygon* is one which is both equiangular and equilateral.

An equilateral triangle is a regular polygon of three sides; a square is one of four.

PROPOSITION I. THEOREM.

Regular polygons of the same number of sides are similar figures.

Let ABCDEF, *abcdef* be two regular polygons of the same number of sides; then will they be similar figures.

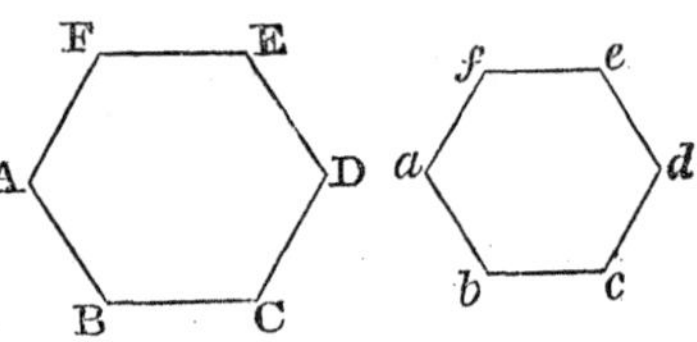

For, since the two polygons have the same number of sides, they must have the same number of angles. Moreover, the sum of the angles of the one polygon is equal to the sum of the angles of the other (Prop. XXVIII., B. I.); and since the polygons are each equiangular, it follows that the angle A is the same part of the sum of the angles A, B, C, D, E, F, that the angle *a* is of the sum of the angles *a, b, c, d, e, f.* Therefore the two angles A and *a* are equal to each other. The same is true of the angles B and *b*, C and *c*, &c.

Moreover, since the polygons are regular, the sides AB, BC, CD, &c., are equal to each other (Def.); so, also, are the sides *ab, bc, cd*, &c. Therefore AB : *ab* : : BC : *bc* : : CD : *cd*, &c. Hence the two polygons have their angles equal, and their homologous sides proportional; they are consequently similar (Def. 3, B. IV.). Therefore, regular polygons, &c.

Cor. The perimeters of two regular polygons of the same number of sides, are to each other as their homologous sides.

and their areas are as the squares of those sides (Prop. XXVI., B. IV.).

Scholium. The angles of a regular polygon are determined by the number of its sides.

PROPOSITION II. THEOREM.

A circle may be described about any regular polygon, and another may be inscribed within it.

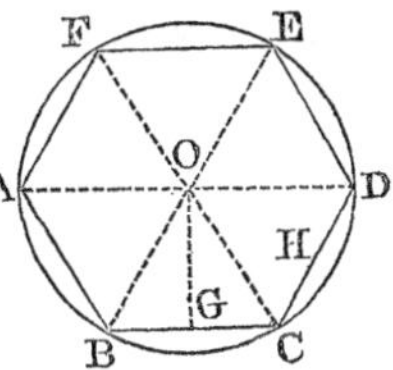

Let ABCDEF be any regular polygon; a circle may be described about it, and another may be inscribed within it.

Bisect the angles FAB, ABC by the straight lines AO, BO; and from the point O in which they meet, draw the lines OC, OD, OE, OF to the other angles of the polygon.

Then, because in the triangles OBA, OBC, AB is, by hypothesis, equal to BC, BO is common to the two triangles, and the included angles OBA, OBC are, by construction, equal to each other; therefore the angle OAB is equal to the angle OCB. But OAB is, by construction, the half of FAB; and FAB is, by hypothesis, equal to DCB; therefore OCB is the half of DCB; that is, the angle BCD is bisected by the line OC. In the same manner it may be proved that the angles CDE, DEF, EFA are bisected by the straight lines OD, OE, OF.

Now because the angles OAB, OBA, being halves of equal angles, are equal to each other, OA is equal to OB (Prop. XI., B. I.). For the same reason, OC, OD, OE, OF are each of them equal to OA. Therefore a circumference described from the center O, with a radius equal to OA, will pass through each of the points B, C, D, E, F, and be described about the polygon.

Secondly. A circle may be inscribed within the polygon ABCDEF. For the sides AB, BC, CD, &c., are equal chords of the same circle; hence they are equally distant from the center O (Prop. VIII., B. III.); that is, the perpendiculars OG, OH, &c., are all equal to each other. Therefore, if from O as a center, with a radius OG, a circumference be described, it will touch the side BC (Prop. IX., B. III.), and each of the other sides of the polygon; hence the circle will be inscribed within the polygon. Therefore a circle may be described, &c.

Scholium 1. In regular polygons, the center of the inscribed and circumscribed circles, is also called the center of the polygon; and the perpendicular from the center upon one of the sides, that is, the radius of the inscribed circle, is called the *apothem* of the polygon.

Since all the chords AB, BC, &c., are equal, the angles at the center, AOB, BOC, &c., are equal; and the value of each may be found by dividing four right angles by the number of sides of the polygon.

Scholium 2. To inscribe a regular polygon of any number of sides in a circle, it is only necessary to divide the circumference into the same number of equal parts; for, if the arcs are equal, the chords AB, BC, CD, &c., will be equal. Hence the triangles AOB, BOC, COD, &c., will also be equal, because they are mutually equilateral; therefore all the angles ABC, BCD, CDE, &c., will be equal, and the figure ABCDEF will be a regular polygon.

PROPOSITION III. PROBLEM.

To inscribe a square in a given circle.

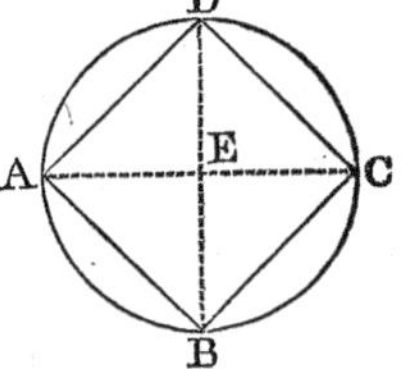

Let ABCD be the given circle; it is required to inscribe a square in it.

Draw two diameters AC, BD at right angles to each other; and join AB, BC, CD, DA. Because the angles AEB, BEC, &c., are equal, the chords AB, BC, &c., are also equal. And because the angles ABC, BCD, &c., are inscribed in semicircles, they are right angles (Prop. XV., Cor. 2, B. III.). Therefore ABCD is a square, and it is inscribed in the circle ABCD.

Cor. Since the triangle AEB is right-angled and isosceles, we have the proportion, AB : AE : : $\sqrt{2}$: 1 (Prop. XI., Cor. 3, B. IV.); therefore *the side of the inscribed square is to the radius, as the square root of* 2 *is to unity.*

PROPOSITION IV. THEOREM.

The side of a regular hexagon is equal to the radius of the circumscribed circle.

Let ABCDEF be a regular hexagon inscribed in a circle whose center is O; then any side as AB will be equal to the radius AO.

Draw the radius BO. Then the angle AOB is the sixth part of four right angles (Prop. II., Sch. 1), or the third part of two right angles. Also, because the three angles of every triangle are equal to two right angles, the two angles OAB, OBA are together equal to two thirds of two right angles; and since AO is equal to BO, each of these angles is one third of two right angles. Hence the triangle AOB is equiangular, and AB is equal to AO. Therefore the side of a regular hexagon, &c.

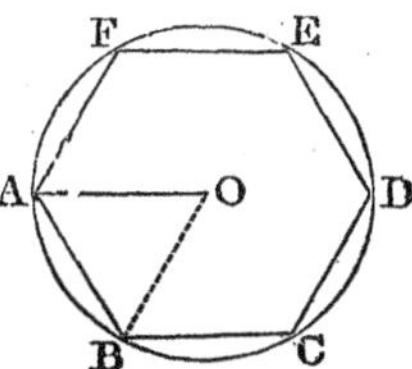

Cor. To inscribe a regular hexagon in a given circle, the radius must be applied six times upon the circumference. By joining the alternate angles A, C, E, an equilateral triangle will be inscribed in the circle.

PROPOSITION V. PROBLEM.

To inscribe a regular decagon in a given circle.

Let ABF be the given circle; it is required to inscribe in it a regular decagon.

Take C the center of the circle; draw the radius AC, and divide it in extreme and mean ratio (Prob. XX., B. V.) at the point D. Make the chord AB equal to CD the greater segment; then will AB be the side of a regular decagon inscribed in the circle.

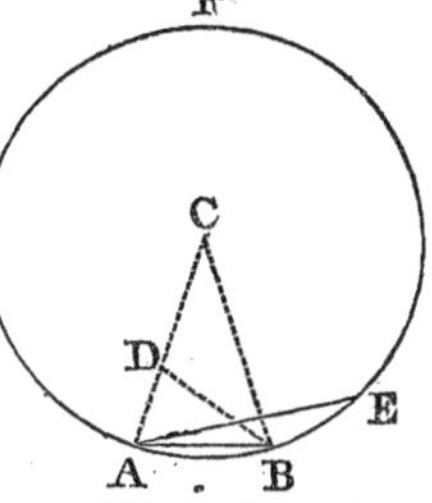

Join BC, BD. Then, by construction, AC : CD : : CD : AD; but AB is equal to CD; therefore AC : AB : : AB : AD. Hence the triangles ACB, ABD have a common angle A included between proportional sides; they are therefore similar (Prop. XX., B. IV.) And because the triangle ACB is isosceles, the triangle ABD must also be isosceles, and AB is equal to BD. But AB was made equal to CD; hence BD is equal to CD, and the angle DBC is equal to the angle DCB. Therefore the exterior angle ADB, which is equal to the sum of DCB and DBC, must be double of DCB. But the angle ADB is equal to DAB; therefore each of the angles CAB, CBA is double of the angle ACB. Hence the sum of the three angles of the triangle ACB is five times the angle C. But these three angles are equal to two right angles (Prop. XXVII., B. I.); therefore the angle C is the fifth part of two right angles, or the tenth part of four right angles. Hence the arc AB is one tenth of the circumference, and the chord AB is the side of a regular decagon inscribed in the circle.

Cor. 1. By joining the alternate angles of the regular decagon, a regular pentagon may be inscribed in the circle.

Cor. 2. By combining this Proposition with the preceding a regular pentedecagon may be inscribed in a circle.

For, let AE be the side of a regular hexagon; then the arc AE will be one sixth of the whole circumference, and the arc AB one tenth of the whole circumference. Hence the arc BE will be $\frac{1}{6}-\frac{1}{10}$ or $\frac{1}{15}$, and the chord of this arc will be the side of a regular pentedecagon.

Scholium. By bisecting the arcs subtended by the sides of any polygon, another polygon of double the number of sides may be inscribed in a circle. Hence the square will enable us to inscribe regular polygons of 8, 16, 32, &c., sides; the hexagon will enable us to inscribe polygons of 12, 24, &c., sides; the decagon will enable us to inscribe polygons of 20, 40, &c., sides; and the pentedecagon, polygons of 30, 60, &c., sides.

The ancient geometricians were unacquainted with any method of inscribing in a circle, regular polygons of 7, 9, 11, 13, 14, 17, &c., sides; and for a long time it was believed that these polygons could not be constructed geometrically; but Gauss, a German mathematician, has shown that a regular polygon of 17 sides may be inscribed in a circle, by employing straight lines and circles only.

PROPOSITION VI. PROBLEM.

A regular polygon inscribed in a circle being given, to describe a similar polygon about the circle.

Let ABCDEF be a regular polygon inscribed in the circle ABD; it is required to describe a similar polygon about the circle.

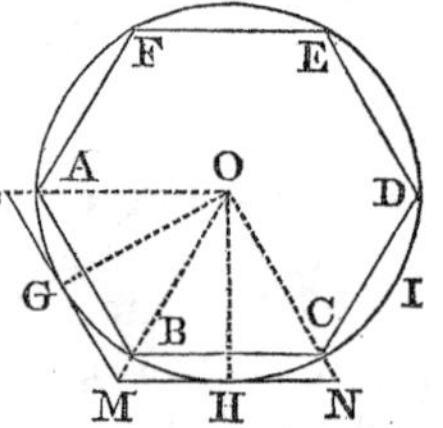

Bisect the arc AB in G, and through G draw the tangent LM. Bisect also the arc BC in H, and through H draw the tangent MN, and in the same manner draw tangents to the middle points of the arcs CD, DE, &c. These tangents, by their intersections, will form a circumscribed polygon similar to the one inscribed.

Find O the center of the circle, and draw the radii OG OH. Then, because OG is perpendicular to the tangent LM (Prop. IX., B. III.), and also to the chord AB (Prop. VI

Sch., B. III.), the tangent is parallel to the chord (Prop. XX., B. I.). In the same manner it may be proved that the other sides of the circumscribed polygon are parallel to the sides of the inscribed polygon; and therefore the angles of the circumscribed polygon are equal to those of the inscribed one (Prop. XXVI., B. I.).

Since the arcs BG, BH are halves of the equal arcs AGB, BHC, they are equal to each other; that is, the vertex B is at the middle point of the arc GBH. Join OM; the line OM will pass through the point B. For the right-angled triangles OMH, OMG have the hypothenuse OM common, and the side OH equal to OG; therefore the angle GOM is equal to the angle HOM (Prop. XIX., B. I.), and the line OM passes through the point B, the middle of the arc GBH.

Now because the triangle OAB is similar to the triangle OLM, and the triangle OBC to the triangle OMN, we have the proportions

AB : LM : : BO : MO;

also, BC : MN : : BO : MO;

therefore (Prop. IV., B. II.),

AB : LM : : BC : MN.

But AB is equal to BC; therefore LM is equal to MN. In the same manner, it may be proved that the other sides of the circumscribed polygon are equal to each other. Hence this polygon is regular, and similar to the one inscribed.

Cor. 1. Conversely, if the circumscribed polygon is given, and it is required to form the similar inscribed one, draw the lines OL, OM, ON, &c., to the angles of the polygon; these lines will meet the circumference in the points A, B, C, &c. Join these points by the lines AB, BC, CD, &c., and a similar polygon will be inscribed in the circle.

Or we may simply join the points of contact G, H, I, &c., by the chords GH, HI, &c., and there will be formed an inscribed polygon similar to the circumscribed one.

Cor. 2. Hence we can circumscribe about a circle, any regular polygon which can be inscribed within it, and conversely.

Cor. 3. A side of the circumscribed polygon MN is equal to twice MH, or MG+MH.

PROPOSITION VII. THEOREM.

The area of a regular polygon is equivalent to the product of its perimeter, by half the radius of the inscribed circle.

Let ABCDEF be a regular polygon, and G the center of

the inscribed circle. From G draw lines to all the angles of the polygon. The polygon will thus be divided into as many triangles as it has sides; and the common altitude of these triangles is GH, the radius of the circle. Now, the area of the triangle BGC is equal to the product of BC by the half of GH (Prop. VI., B. IV.); and so of all the other triangles having their vertices in G. Hence the sum of all the triangles, that is, the surface of the polygon, is equivalent to the product of the sum of the bases AB, BC, &c.; that is, the perimeter of the polygon, multiplied by half of GH, or half the radius of the inscribed circle. Therefore, the area of a regular polygon, &c.

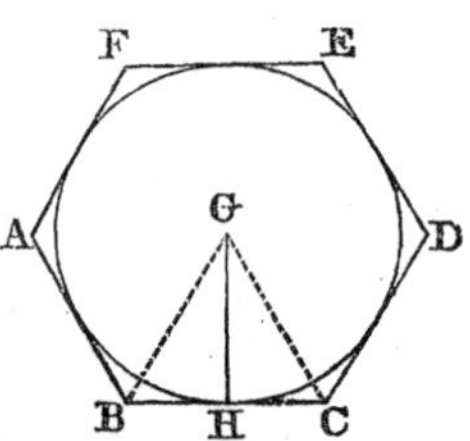

PROPOSITION VIII. THEOREM.

The perimeters of two regular polygons of the same number of sides, are as the radii of the inscribed or circumscribed circles, and their surfaces are as the squares of the radii.

Let ABCDEF, *abcdef* be two regular polygons of the same number of sides; let G and *g* be the centers of the circumscribed circles; and let GH, *gh* be drawn perpendicular to BC and *bc*; then will the perimeters of the polygons be as the radii BG, *bg*; and, also, as GH, *gh*, the radii of the inscribed circles.

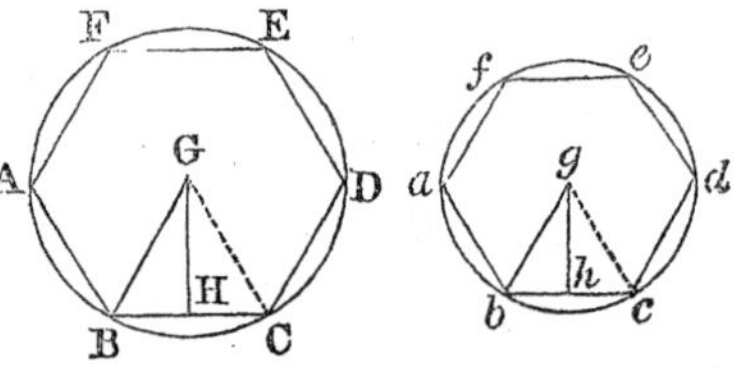

The angle BGC is equal to the angle *bgc* (Prop. II., Sch. 1); and since the triangles BGC, *bgc* are isosceles, they are similar. So, also, are the right-angled triangles BGH, *bgh*; and, consequently, BC : *bc* : : BG : *bg* : : GH : *gh*. But the perimeters of the two polygons are to each other as the sides BC, *bc* (Prop. I., Cor.); they are, therefore, to each other as the radii BG, *bg* of the circumscribed circles; and also as the radii GH, *gh* of the inscribed circles.

The surfaces of these polygons are to each other as the squares of the homologous sides BC, *bc* (Prop. I., Cor.); they are, therefore, as the squares of BG, *bg*, the radii of the circumscribed circles; or as the squares of GH, *gh*, the radii of the inscribed circles.

PROPOSITION IX. PROBLEM.

The surface of a regular inscribed polygon, and that of a similar circumscribed polygon, being given; to find the surfaces of regular inscribed and circumscribed polygons having double the number of sides.

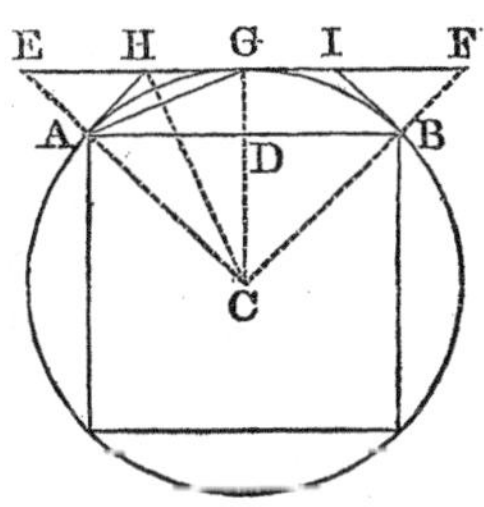

Let AB be a side of the given inscribed polygon; EF parallel to AB, a side of the similar circumscribed polygon; and C the center of the circle. Draw the chord AG, and it will be the side of the inscribed polygon having double the number of sides. At the points A and B draw tangents, meeting EF in the points H and I; then will HI, which is double of HG, be a side of the similar circumscribed polygon (Prop. VI., Cor. 1). Let p represent the inscribed polygon whose side is AB, P the corresponding circumscribed polygon; p' the inscribed polygon having double the number of sides, P′ the similar circumscribed polygon. Then it is plain that the space CAD is the same part of p, that CEG is of P; also, CAG of p', and CAHG of P′; for each of these spaces must be repeated the same number of times, to complete the polygons to which they severally belong.

First. The triangles ACD, ACG, whose common vertex is A, are to each other as their bases CD, CG; they are also to each other as the polygons p and p'; hence

$$p : p' :: CD : CG.$$

Again, the triangles CGA, CGE, whose common vertex is G are to each other as their bases CA, CE; they are also to each other as the polygons p' and P; hence

$$p' : P :: CA : CE.$$

But since AD is parallel to EG, we have CD : CG : : CA CE; therefore,

$$p \quad p' :: p' : P;$$

that is, *the polygon* p′ *is a mean proportional between the two given polygons.*

Secondly. The triangles CGH, CHE, having the common altitude CG, are to each other as their bases GH, HE. But since CH bisects the angle GCE, we have (Prop. XVII., B. IV.),

$$GH : HE :: CG : CE :: CD : CA, \text{ or } CG \quad : p : p'.$$

Therefore, $\quad CGH : CHE :: p : p';$

hence (Prop. VI., B. II.)

$$CGH : CGH+CHE, \text{ or } CGE :: p : p+p',$$

or $$2CGH : CGE :: 2p : p+p'.$$

But $$2CGH, \text{ or } CGHA : CGE :: P' : P.$$

Therefore, $P' : P :: 2p : p+p'$; whence $P'=\dfrac{2pP}{p+p'}$;

that is, *the polygon* P′ *is found by dividing twice the product of the two given polygons by the sum of the two inscribed polygons*

Hence, by means of the polygons p and P, it is easy to find the polygons p' and P′ having double the number of sides.

PROPOSITION X. THEOREM.

A circle being given, two similar polygons can always be found, the one described about the circle, and the other inscribed in it, which shall differ from each other by less than any assignable surface.

Let ACD be the given circle, and the square of X any given surface; a polygon can be inscribed in the circle ACD, and a similar polygon be described about it, such that the difference between them shall be less than the square of X.

Bisect AC a fourth part of the circumference, then bisect the half of this fourth, and so continue the bisection, until an arc is found whose chord AB is less than X. As this arc must be contained a certain number of times exactly in the whole circumference, if we apply chords AB, BC, &c., each equal to AB, the last will terminate at A, and a regular polygon ABCD, &c., will be inscribed in the circle.

Next describe a similar polygon about the circle (Prop. VI.); the difference of these two polygons will be less than the square of X.

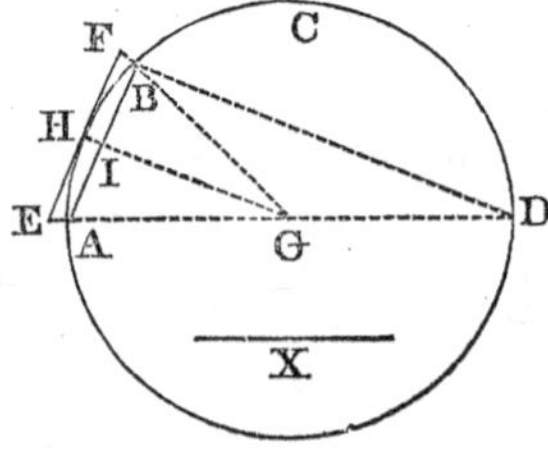

Find the center G, and draw the diameter AD. Let EF be a side of the circumscribed polygon; and join EG, FG. These lines will pass through the points A and B, as was shown in Prop. VI. Draw GH to the point of contact H; it will bisect AB in I, and be perpendicular to it (Prop. VI., Sch., B. III.). Join, also, BD.

Let P represent the circumscribed polygon, and p the inscribed polygon. Then, because the polygons are similar, they are as the squares of the homologous sides EF and AB

(Prop. XXVI., B. IV.); that is, because the triangles EFG ABG are similar, as the square of EG to the square of AG that is, of HG.

Again, the triangles EHG, ABD, having their sides parallel to each other, are similar; and, therefore,

$$EG : HG :: AD : BD.$$

But the polygon P is to the polygon p as the square of EG to the square of HG;

hence $P : p :: AD^2 : BD^2$,

and, by division, $P : P-p :: AD^2 : AD^2-BD^2$, or AB^2.

But the square of AD is greater than a regular polygon of eight sides described about the circle, because it contains that polygon; and for the same reason, the polygon of eight sides is greater than the polygon of sixteen, and so on. Therefore P is less than the square of AD; and, consequently (Def. 2, B. II.), $P-p$ is less than the square of AB; that is, less than the given square on X. Hence, the difference of the two polygons is less than the given surface.

Cor. Since the circle can not be less than any inscribed polygon, nor greater than any circumscribed one, it follows that *a polygon may be inscribed in a circle, and another described about it, each of which shall differ from the circle by less than any assignable surface.*

PROPOSITION XI. PROBLEM.

To find the area of a circle whose radius is unity.

If the radius of a circle be unity, the diameter will be represented by 2, and the area of the circumscribed square will be 4; while that of the inscribed square, being half the circumscribed, is 2. Now, according to Prop. IX., the surface of the inscribed octagon, is a mean proportional between the two squares p and P, so that $p'=\sqrt{8}=2.82843$. Also, the circumscribed octagon $P'=\frac{2pP}{p+p'}=\frac{16}{2+\sqrt{8}}=3.31371$. Having thus obtained the inscribed and circumscribed octagons, we may in the same way determine the polygons having twice the number of sides. We must put $p=2.82843$, and $P=3.31371$, and we shall have $p'=\sqrt{pP}=3.06147$; and $P'=\frac{2pP}{p+p'}=3.18260$. These polygons of 16 sides will furnish us those of 32; and thus we may proceed, until there is no difference between the inscribed and circumscribed polygons, at least for any number of decimal places which may be desired. The following table gives the results of this computation for five decimal places:

Number of Sides.	Inscribed Polygon.	Circumscribed Polygon
4	2.00000	4.00000
8	2.82843	3.31371
16	3.06147	3.18260
32	3.12145	3.15172
64	3.13655	3.14412
128	3.14033	3.14222
256	3.14128	3.14175
512	3.14151	3.14163
1024	3.14157	3.14160
2048	3.14159	3.14159

Now as the inscribed polygon can not be greater than the circle, and the circumscribed polygon can not be less than the circle, it is plain that 3.14159 must express the area of a circle, whose radius is unity, correct to five decimal places.

After three bisections of a quadrant of a circle, we obtain the inscribed polygon of 32 sides, which differs from the corresponding circumscribed polygon, only in the second decimal place. After five bisections, we obtain polygons of 128 sides, which differ only in the third decimal place; after nine bisections, they agree to five decimal places, but differ in the sixth place; after eighteen bisections, they agree to ten decimal places; and thus, by continually bisecting the arcs subtended by the sides of the polygon, new polygons are formed, both inscribed and circumscribed, which agree to a greater number of decimal places. Vieta, by means of inscribed and circumscribed polygons, carried the approximation to ten places of figures; Van Ceulen carried it to 36 places; Sharp computed the area to 72 places; De Lagny to 128 places; and Dr. Clausen has carried the computation to 250 places of decimals.

By continuing this process of bisection, the difference between the inscribed and circumscribed polygons may be made less than any quantity we can assign, however small. The number of sides of such a polygon will be indefinitely great; and hence a regular polygon of an infinite number of sides, is said to be ultimately equal to the circle. Henceforth, we shall therefore regard the circle as a regular polygon of an infinite number of sides.

PROPOSITION XII. THEOREM.

The area of a circle is equal to the product of its circumference by half the radius.

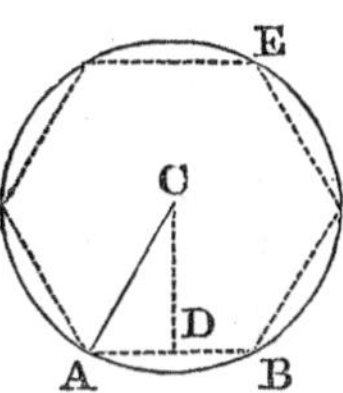

Let ABE be a circle whose center is C and radius CA; the area of the circle is equal to the product of its circumference by half of CA.

Inscribe in the circle any regular polygon, and from the center draw CD perpendicular to one of the sides. The area of the polygon will be equal to its perimeter multiplied by half of CD (Prop. VII.). Conceive the number of sides of the polygon to be indefinitely increased, by continually bisecting the arcs subtended by the sides; its perimeter will ultimately coincide with the circumference of the circle the perpendicular CD will become equal to the radius CA and the area of the polygon to the area of the circle (Prop XI.). Consequently, the area of the circle is equal to the product of its circumference by half the radius.

Cor. The area of a sector is equal to the product of its arc by half its radius.

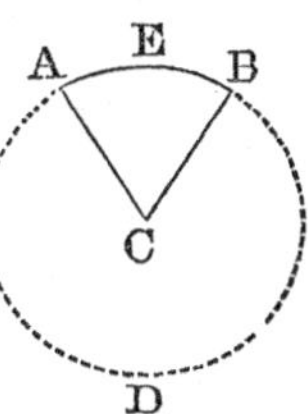

For the sector ACB is to the whole circle ABD, as the arc AEB is to the whole circumference ABD (Prop. XIV., Cor. 2, B. III.); or, since magnitudes have the same ratio which their equimultiples have (Prop. VIII., B. II.), as the arc AEB $\times \frac{1}{2}$AC is to the circumference ABD $\times \frac{1}{2}$AC. But this last expression is equal to the area of the circle; therefore the area of the sector ACB is equal to the product of its arc AEB by half of AC.

PROPOSITION XIII. THEOREM.

The circumferences of circles are to each other as their radii, and their areas are as the squares of their radii.

Let R and r denote the radii of two circles; C and c their circumferences; A and a their areas; then we shall have

$$C : c :: R : r,$$

and

$$A : a :: R^2 : r^2$$

Inscribe within the circles, two regular polygons having

the same number of sides. Now whatever be the number of sides of the polygons, their perimeters will be to each other as the radii of the circumscribed circles (Prop. VIII.). Conceive the arcs subtended by the sides of the polygons to be continually bisected, until the number of sides of the polygons becomes indefinitely great, the perimeters of the polygons will ultimately become equal to the circumferences of the circles, and we shall have

$$C : c :: R : r.$$

Again, the areas of the polygons are to each other as the squares of the radii of the circumscribed circles (Prop. VIII.). But when the number of sides of the polygons is indefinitely increased, the areas of the polygons become equal to the areas of the circles, and we shall have

$$A : a :: R^2 : r^2.$$

Cor. 1. Similar arcs are to each other as their radii; and similar sectors are as the squares of their radii.

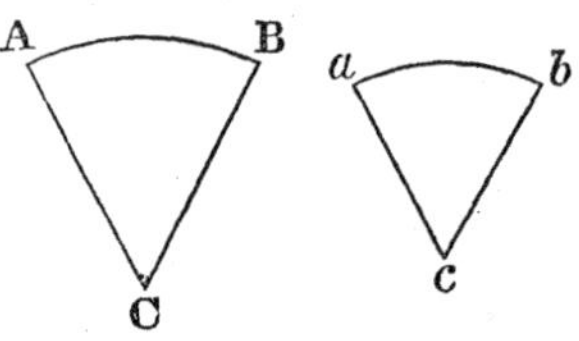

For since the arcs AB, *ab* are similar, the angle C is equal to the angle *c* (Def. 5, B. IV.). But the angle C is to four right angles, as the arc AB is to the whole circumference described with the radius AC (Prop. XIV., B. III.); and the angle *c* is to four right angles, as the arc *ab* is to the circumference described with the radius *ac*. Therefore the arcs AB, *ab* are to each other as the circumferences of which they form a part. But these circumferences are to each other as AC, *ac*; therefore,

$$\textit{Arc } AB : \textit{arc } ab :: AC : ac.$$

For the same reason, the sectors ACB, *acb* are as the entire circles to which they belong; and these are as the squares of their radii; therefore,

$$\textit{Sector } ACB : \textit{sector } acb :: AC^2 : ac^2.$$

Cor. 2. Let π represent the circumference of a circle whose diameter is unity; also, let D represent the diameter, R the radius, and C the circumference of any other circle; then, since the circumferences of circles are to each other as their diameters,

$$1 : \pi :: 2R : C;$$

therefore, $$C = 2\pi R = \pi D;$$

that is, *the circumference of a circle is equal to the product of its diameter by the constant number* π.

Cor. 3. According to Prop. XII., the area of a circle is equal to the product of its circumference by half the radius

If we put A to represent the area of a circle, then

$$A = C \times \tfrac{1}{2}R = 2\pi R \times \tfrac{1}{2}R = \pi R^2;$$

that is, *the area of a circle is equal to the product of the square of its radius by the constant number* π.

Cor. 4. When R is equal to unity, we have $A = \pi$; that is, π *is equal to the area of a circle whose radius is unity.* According to Prop. XI., π is therefore equal to 3.14159 nearly This number is represented by π, because it is the first letter of the Greek word which signifies circumference.

SOLID GEOMETRY.

BOOK VII.

PLANES AND SOLID ANGLES

Definitions.

1. A STRAIGHT line is *perpendicular to a plane,* when it is perpendicular to every straight line which it meets in that plane.

Conversely, the plane in this case is perpendicular to the line.

The *foot* of the perpendicular, is the point in which it meets the plane.

2. A line is *parallel to a plane,* when it can not meet the plane, though produced ever so far.

Conversely, the plane in this case is parallel to the line.

3. Two *planes are parallel* to each other, when they can not meet, though produced ever so far.

4. The *angle contained by two planes* which cut each other, is the angle contained by two lines drawn from any point in the line of their common section, at right angles to that line, one in each of the planes.

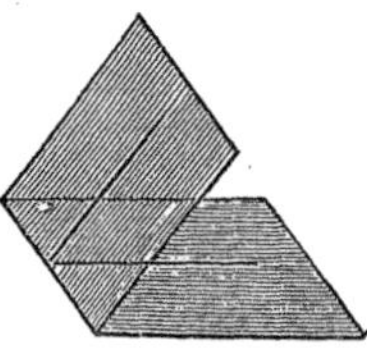

This angle may be acute, right, or obtuse.

If it is a right angle, the two planes are perpendicular to each other.

5. A *solid angle* is the angular space contained by more than two planes which meet at the same point.

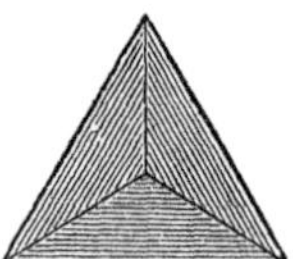

PROPOSITION I. THEOREM

One part of a straight line can not be in a plane, and another part without it.

For from the definition of a plane (Def. 6, B. I.), when a

straight line has two points common with a plane it lies wholly in that plane.

Scholium. To discover whether a surface is plane, we apply a straight line in different directions to this surface, and see if it touches throughout its whole extent.

PROPOSITION II. THEOREM.

Any two straight lines which cut each other, are in one plane, and determine its position.

Let the two straight lines AB, BC cut each other in B; then will AB, BC be in the same plane.

Conceive a plane to pass through the straight line BC, and let this plane be turned about BC, until it pass through the point A. Then, because the points A and B are situated in this plane the straight line AB lies in it (Def. 6, B. I.). Hence the position of the plane is determined by the condition of its containing the two lines AB, BC. Therefore, any two straight lines, &c.

Cor. 1. A triangle ABC, or three points A, B, C, not in the same straight line, determine the position of a plane.

Cor. 2. Two parallel lines AB, CD determine the position of a plane. For if the line EF be drawn, the plane of the two straight lines AE, EF will be the same as that of the parallels AB, CD; and it has already been proved that two straight lines which cut each other, determine the position of a plane.

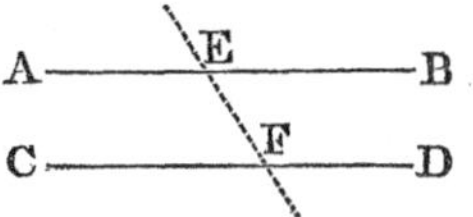

PROPOSITION III. THEOREM.

If two planes cut each other, their common section is a straight line.

Let the two planes AB, CD cut each other, and let E. F be two points in their common section. From E to F draw the straight line EF. Then, since the points E and F are in the plane AB, the straight line EF which joins them, must lie wholly in that plane (Def. 6, B. I.). For the same reason, EF must lie wholly in the plane

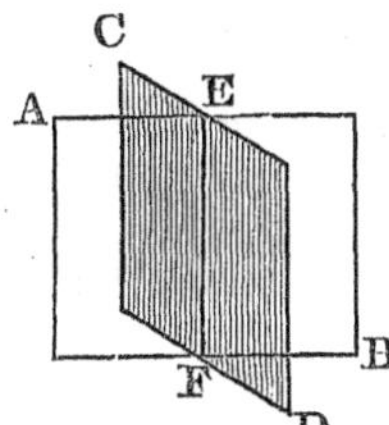

CD. Therefore the straight line EF is common to the two planes AB, CD; that is, it is their common section. Hence, if two planes, &c.

PROPOSITION IV. THEOREM.

If a straight line be perpendicular to each of two straight lines at their point of intersection, it will be perpendicular to the plane in which these lines are.

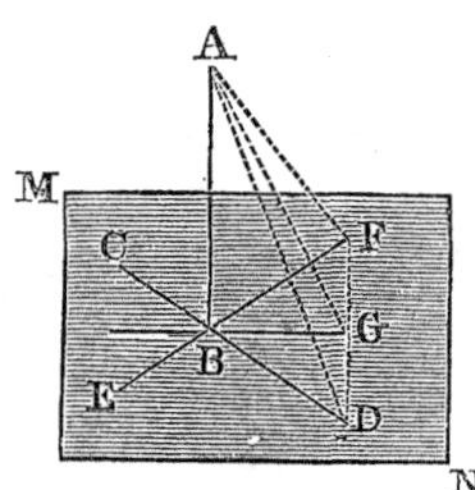

Let the straight line AB be perpendicular to each of the straight lines CD, EF which intersect at B; AB will also be perpendicular to the plane MN which passes through these lines.

Through B draw any line BG, in the plane MN; let G be any point of this line, and through G draw DGF, so that DG shall be equal to GF (Prob. XXI., B. V.). Join AD, AG, and AF.

Then, since the base DF of the triangle DBF is bisected in G, we shall have (Prop. XIV., B. IV.),

$$BD^2+BF^2=2BG^2+2GF^2.$$

Also, in the triangle DAF,

$$AD^2+AF^2=2AG^2+2GF^2.$$

Subtracting the first equation from the second, we have

$$AD^2-BD^2+AF^2-BF^2=2AG^2-2BG^2.$$

But, because ABD is a right-angled triangle,

$$AD^2-BD^2=AB^2;$$

and, because ABF is a right-angled triangle,

$$AF^2-BF^2=AB^2.$$

Therefore, substituting these values in the former equation,

$$AB^2+AB^2=2AG^2-2BG^2;$$

whence $$AB^2=AG^2-BG^2,$$

or $$AG^2=AB^2+BG^2.$$

Wherefore ABG is a right angle (Prop. XIII., Sch., B. IV.); that is, AB is perpendicular to the straight line BG. In like manner, it may be proved that AB is perpendicular to any other straight line passing through B in the plane MN; hence it is perpendicular to the plane MN (Def. 1). Therefore, if a straight line, &c.

Scholium. Hence it appears not only that a straight line *may be* perpendicular to every straight line which passes through its foot in a plane, but that it always *must be* so whenever it is perpendicular to two lines in the plane, which shows that the first definition involves no impossibility.

Cor. 1. The perpendicular AB is shorter than any oblique line AD; it therefore measures the true distance of the point A from the plane MN.

Cor. 2. Through a given point B in a plane, only one perpendicular can be drawn to this plane. For, if there could be two perpendiculars, suppose a plane to pass through them, whose intersection with the plane MN is BG; then these two perpendiculars would both be at right angles to the line BG, at the same point and in the same plane, which is impossible (Prop. XVI., Cor., B. I.).

It is also impossible, from a given point without a plane, to let fall two perpendiculars upon the plane. For, suppose AB, AG to be two such perpendiculars; then the triangle ABG will have two right angles, which is impossible (Prop. XXVII., Cor. 3, B. I.).

PROPOSITION V. THEOREM.

Oblique lines drawn from a point to a plane, at equal distances from the perpendicular, are equal; and of two oblique lines unequally distant from the perpendicular, the more remote is the longer.

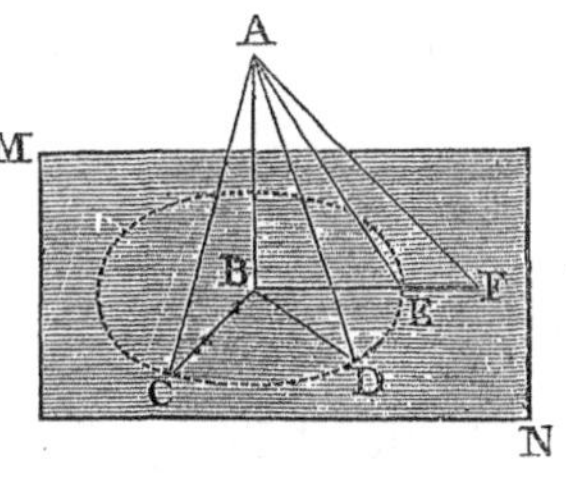

Let the straight line AB be drawn perpendicular to the plane MN; and let AC, AD, AE be oblique lines drawn from the point A, equally distant from the perpendicular; also, let AF be more remote from the perpendicular than AE; then will the lines AC, AD, AE all be equal to each other, and AF be longer than AE.

For, since the angles ABC, ABD, ABE are right angles and BC, BD, BE are equal, the triangles ABC, ABD, ABE have two sides and the included angle equal; therefore the third sides AC, AD, AE are equal to each other.

So, also, since the distance BF is greater than BE, it is plain that the oblique line AF is longer than AE (Prop. XVII., B. I.).

Cor. All the equal oblique lines AC, AD, AE, &c., terminate in the circumference CDE, which is described from B, the foot of the perpendicular, as a center.

If, then, it is required to draw a straight line perpendicular to the plane MN, from a point A without it, take three points in the plane C, D, E, equally distant from A, and find B the center of the circle which passes througn these points. Join AB, and it will be the perpendicular required.

Scholium. The angle AEB is called *the inclination of the line* AE to the plane MN. All the lines AC, AD, AE, &c., which are equally distant from the perpendicular, have the same inclination to the plane; because all the angles ACB, ADB, AEB, &c., are equal.

PROPOSITION VI. THEOREM.

If a straight line is perpendicular to a plane, every plane which passes through that line, is perpendicular to the first-mentioned plane.

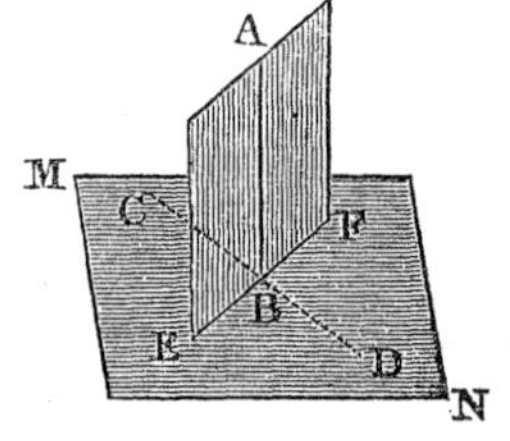

Let the straight line AB be perpendicular to the plane MN; then will every plane which passes through AB be perpendicular to the plane MN.

Suppose any plane, as AE, to pass through AB, and let EF be the common section of the planes AE, MN. In the plane MN, through the point B, draw CD perpendicular to the common section EF. Then, since the line AB is perpendicular to the plane MN, it must be perpendicular to each of the two straight lines CD, EF (Def. 1). But the angle ABD, formed by the two perpendiculars BA, BD, to the common section EF, measures the angle of the two planes AE, MN (Def. 4); and since this is a right angle, the two planes must be perpendicular to each other. Therefore, if a straight line, &c.

Scholium. When three straight lines, as AB, CD, EF, are perpendicular to each other, each of these lines is perpendicular to the plane of the other two, and the three planes are perpendicular to each other.

PROPOSITION VII THEOREM.

If two planes are perpendicular to each other, a straight line drawn in one of them perpendicular to their common section, will be perpendicular to the other plane.

Let the plane AE be perpendicular to the plane MN, and let the line AB be drawn in the plane AE perpendicular to the common section EF; then will AB be perpendicular to the plane MN.

For in the plane MN, draw CD through the point B perpendicular to EF. Then, because the planes AE and MN are perpendicular, the angle ABD is a right angle. Hence the line AB is perpendicular to the two straight lines CD, EF at their point of intersection; it is consequently perpendicular to their plane MN (Prop. IV.). Therefore, if two planes, &c.

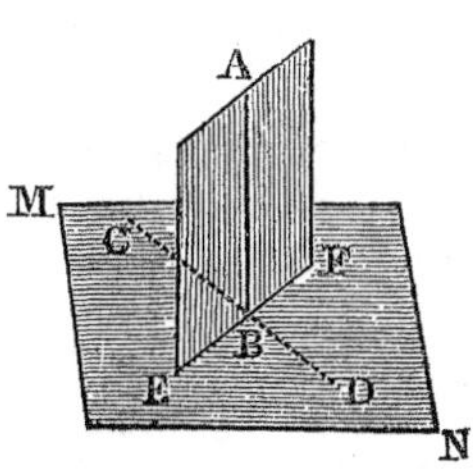

Cor. If the plane AE is perpendicular to the plane MN, and if from any point B, in their common section, we erect a perpendicular to the plane MN, this perpendicular will be in the plane AE. For if not, then we may draw from the same point, a straight line AB in the plane AE perpendicular to EF, and this line, according to the Proposition, will be perpendicular to the plane MN. Therefore there would be two perpendiculars to the plane MN, drawn from the same point, which is impossible (Prop. IV., Cor. 2).

PROPOSITION VIII. THEOREM.

If two planes, which cut one another, are each of them perpendicular to a third plane, their common section is perpendicular to the same plane.

Let the two planes AE, AD be each of them perpendicular to a third plane MN, and let AB be the common section of the first two planes; then will AB be perpendicular to the plane MN.

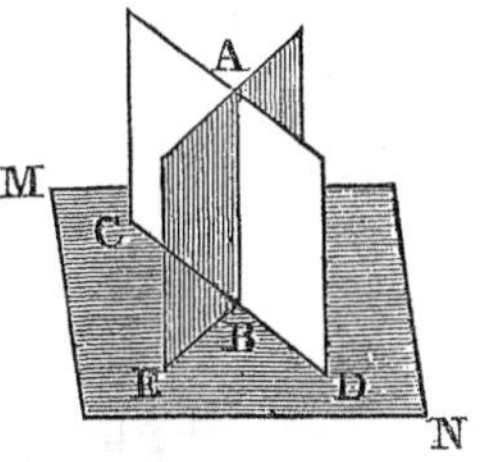

For, from the point B, erect a perpendicular to the plane MN. Then, by the Corollary of the last Proposition, this line must be situated both in the plane AD and in the plane AE; hence it is their common section AB. Therefore, if two planes, &c.

PROPOSITION IX. THEOREM.

Two straight lines which are perpendicular to the same plane, are parallel to each other.

Let the two straight lines AB, CD be each of them perpendicular to the same plane MN; then will AB be parallel to CD

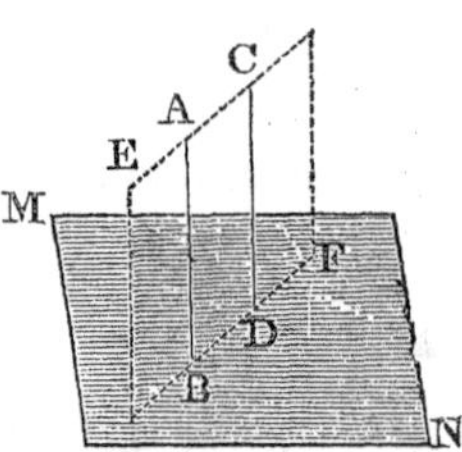

In the plane MN, draw the straight line BD joining the points B and D. Through the lines AB, BD pass the plane EF; it will be perpendicular to the plane MN (Prop. VI.); also, the line CD will lie in this plane, because it is perpendicular to MN (Prop. VII., Cor.). Now, because AB and CD are both perpendicular to the plane MN, they are perpendicular to the line BD in that plane; and since AB, CD are both perpendicular to the same line BD, and lie in the same plane, they are parallel to each other (Prop. XX., B. I.). Therefore, two straight lines, &c.

Cor. 1. If one of two parallel lines be perpendicular to a plane, the other will be perpendicular to the same plane. If AB is perpendicular to the plane MN, then (Prop. VI.) the plane EF will be perpendicular to MN. Also, AB is perpendicular to BD; and if CD is parallel to AB, it will be perpendicular to BD, and therefore (Prop. VII.) it is perpendicular to the plane MN.

Cor. 2. Two straight lines, parallel to a third, are parallel to each other. For, suppose a plane to be drawn perpendicular to any one of them; then the other two, being parallel to the first, will be perpendicular to the same plane, by the preceding Corollary; hence, by the Proposition, they will be parallel to each other.

The three straight lines are supposed not to be in the same plane; for in this case the Proposition has been already demonstrated

PROPOSITION X. THEOREM.

If a straight line, without a given plane, be parallel to a straight line in the plane, it will be parallel to the plane.

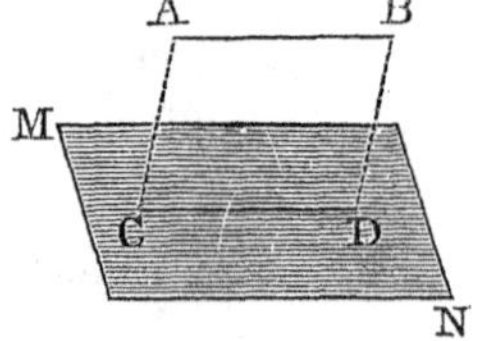

Let the straight line AB be parallel to the straight line CD, in the plane MN; then will it be parallel to the plane MN.

Through the parallels AB, CD suppose a plane ABDC to pass. If the line AB can meet the plane MN, it must meet it in some point of the line CD, which is the common intersection of the two planes. But AB can not meet CD, since they are parallel; hence it can not meet the plane MN that is, AB is parallel to the plane MN (Def. 2). Therefore, if a straight line &c.

PROPOSITION XI. THEOREM.

Two planes, which are perpendicular to the same straight line, are parallel to each other.

Let the planes MN, PQ be perpendicular to the line AB; then will they be parallel to each other.

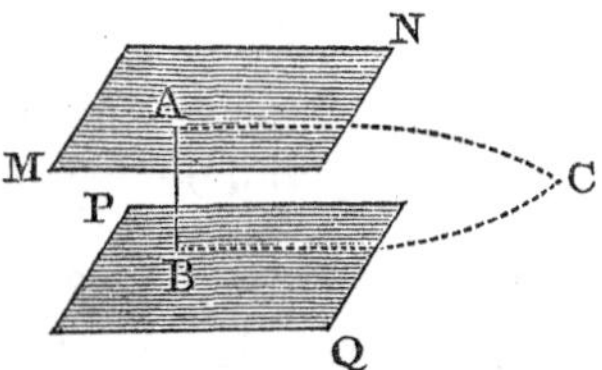

For if they are not parallel, they will meet if produced. Let them be produced and meet in C. Join AC, BC. Now the line AB, which is perpendicular to the plane MN, is perpendicular to the line AC drawn through its foot in that plane. For the same reason AB is perpendicular to BC. Therefore CA and CB are two perpendiculars let fall from the same point C upon the same straight line AB, which is impossible (Prop. XVI., B. I.). Hence the planes MN, PQ can not meet when produced; that is, they are parallel to each other. Therefore, two planes, &c.

PROPOSITION XII. THEOREM.

If two parallel planes are cut by a third plane, their common sections are parallel.

Let the parallel planes MN, PQ be cut by the plane ABDC; and let their common sections with it be AB, CD; then will AB be parallel to CD.

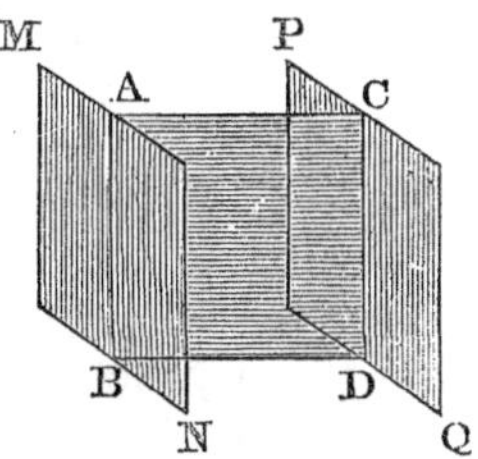

For the two lines AB, CD are in the same plane, viz., in the plane ABDC which cuts the planes MN, PQ; and if these lines were not parallel, they would meet when produced; therefore the planes MN, PQ would also meet, which is impossible, because they are parallel. Hence the lines AB, CD are parallel. Therefore, if two parallel planes, &c.

PROPOSITION XIII. THEOREM.

If two planes are parallel, a straight line which is perpendicular to one of them, is also perpendicular to the other.

Let the two planes MN, PQ be parallel, and let the straight line AB be perpendicular to the plane MN; AB will also be perpendicular to the plane PQ.

Through the point B, draw any line BD in the plane PQ; and through the lines AB, BD suppose a plane to pass intersecting the plane MN in AC. The two lines AC, BD will be parallel (Prop. XII.). But the line AB, being perpendicular to the plane MN, is perpendicular to the straight line AC which it meets in that plane; it must, therefore, be perpendicular to its parallel BD (Prop. XXIII., Cor. 1, B. I.). But BD is any line drawn through B in the plane PQ; and since AB is perpendicular to any line drawn through its foot in the plane PQ, it must be perpendicular to the plane PQ (Def. 1). Therefore, if two planes, &c.

PROPOSITION XIV. THEOREM.

Parallel straight lines included between two parallel planes are equal.

Let AB, CD be the two parallel straight lines included between two parallel planes MN, PQ; then will AB be equal to CD.

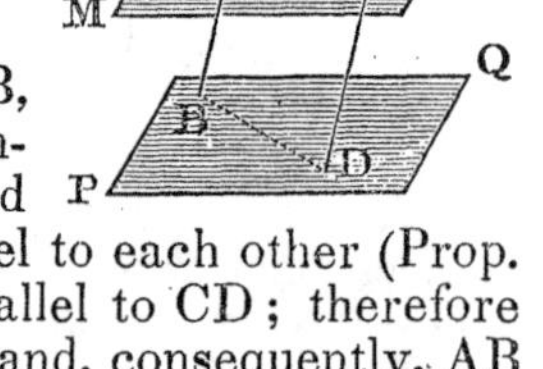

Through the two parallel lines. AB, CD suppose a plane ABDC to pass, intersecting the parallel planes in AC and BD. The lines AC, BD will be parallel to each other (Prop. XII.). But AB is, by supposition, parallel to CD; therefore the figure ABDC is a parallelogram; and, consequently, AB is equal to CD (Prop. XXIX., B. I.). Therefore, parallel straight lines, &c.

Cor. Hence *two parallel planes are every where equidistant;* for if AB, CD are perpendicular to the plane MN, they will be perpendicular to the parallel plane PQ (Prop. XIII.); and being both perpendicular to the same plane, they will be parallel to each other (Prop IX.), and, consequently, equal.

PROPOSITION XV. THEOREM.

If two angles, not in the same plane, have their sides parallel and similarly situated, these angles will be equal, and their planes will be parallel.

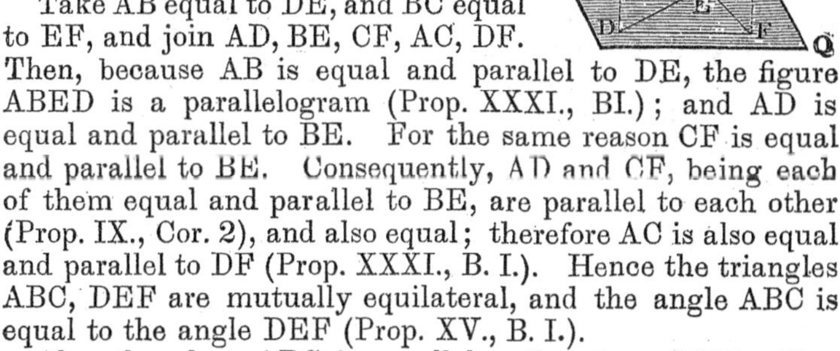

Let the two angles ABC, DEF, lying in different planes MN, PQ, have their sides parallel each to each and similarly situated; then will the angle ABC be equal to the angle DEF, and the plane MN be parallel to the plane PQ.

Take AB equal to DE, and BC equal to EF, and join AD, BE, CF, AC, DF. Then, because AB is equal and parallel to DE, the figure ABED is a parallelogram (Prop. XXXI., BI.); and AD is equal and parallel to BE. For the same reason CF is equal and parallel to BE. Consequently, AD and CF, being each of them equal and parallel to BE, are parallel to each other (Prop. IX., Cor. 2), and also equal; therefore AC is also equal and parallel to DF (Prop. XXXI., B. I.). Hence the triangles ABC, DEF are mutually equilateral, and the angle ABC is equal to the angle DEF (Prop. XV., B. I.).

Also, the plane ABC is parallel to the plane DEF. For, if they are not parallel, suppose a plane to pass through A parallel to DEF, and let it meet the straight lines BE, CF in the points G and H. Then the three lines AD, GE, HF will be equal (Prop. XIV.). But the three lines AD, BE, CF have already been proved to be equal; hence BE is equal to GE, and CF is equal to HF, which is absurd; consequently, the plane ABC must be parallel to the plane DEF. Therefore, if two angles, &c.

Cor. 1. If two parallel planes MN, PQ are met by two other planes ABED, BCFE, the angles formed by the intersections of the parallel planes will be equal. For the section AB is parallel to the section DE (Prop. XII.); and BC is parallel to EF; therefore, by the Proposition, the angle ABC is equal to the angle DEF.

Cor. 2. If three straight lines AD, BE, CF, not situated in the same plane, are equal and parallel, the triangles ABC, DEF, formed by joining the extremities of these lines, will be equal, and their planes will be parallel. For, since AD is equal and parallel to BE, the figure ABED is a parallelogram; hence the side AB is equal and parallel to DE. For the same reason, the sides BC and EF are equal and parallel; as, also, the sides AC and DF. Consequently, the two triangles ABC, DEF are equal; and, according to the Proposition, their planes are parallel.

PROPOSITION XVI. THEOREM.

If two straight lines are cut by parallel planes, they will be cut in the same ratio.

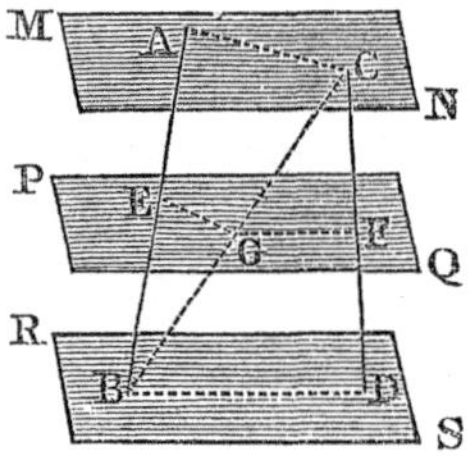

Let the straight lines AB, CD be cut by the parallel planes MN, PQ, RS in the points A, E, B, C, F, D; then we shall have the proportion

AE : EB : : CF : FD.

Draw the line BC meeting the plane PQ in G, and join AC, BD, EG, GF. Then, because the two parallel planes MN, PQ are cut by the plane ABC, the common sections AC, EG are parallel (Prop. XII.). Also, because the two parallel planes PQ, RS are cut by the plane BCD, the common sections BD, GF are parallel. Now, because EG is parallel to AC, a side of the triangle ABC (Prop. XVI., B. IV.), we have

AE : EB : : CG : GB.

Also, because GF is parallel to BD, one side of the triangle BCD, we have

CG : GB : : CF : FD;

hence (Prop. IV., B. II.),

AE : EB : : CF : FD.

Therefore, if two straight lines, &c.

PROPOSITION XVII. THEOREM.

If a solid angle is contained by three plane angles, the sum of any two of these angles is greater than the third.

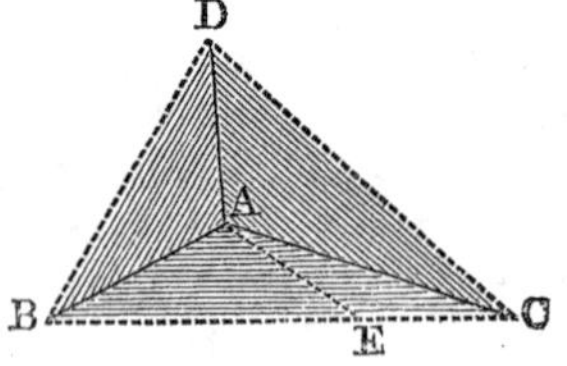

Let the solid angle at A be contained by the three plane angles BAC, CAD, DAB; any two of these angles will be greater than the third.

If these three angles are all equal to each other, it is plain that any two of them must be greater than the third. But if they are not equal, let BAC be that angle which is no less than either of the other two, and is greater than one of them BAD. Then, at the point A, make the angle BAE equal to the angle BAD; take AE equal to AD; through E draw the line BEC cutting AB, AC in the points B and C; and join DB, DC.

Now, because, in the two triangles BAD, BAE, AD is equal to AE, AB is common to both, and the angle BAD is equal to the angle BAE; therefore the base BD is equal to the base BE (Prop. VI., B. I.). Also, because the sum of the lines BD, DC is greater than BC (Prop. VIII., B. I.), and BD is proved equal to BE, a part of BC, therefore the remaining line DC is greater than EC. Now, in the two triangles CAD, CAE, because AD is equal to AE, AC is common, but the base CD is greater than the base CE; therefore the angle CAD is greater than the angle CAE (Prop. XIV., B. I.). But, by construction, the angle BAD is equal to the angle BAE; therefore the two angles BAD, CAD are together greater than BAE, CAE; that is, than the angle BAC. Now BAC is not less than either of the angles BAD, CAD; hence BAC, with either of them, is greater than the third. Therefore, if a solid angle, &c.

PROPOSITION XVIII. THEOREM.

The plane angles which contain any solid angle, are together less than four right angles.

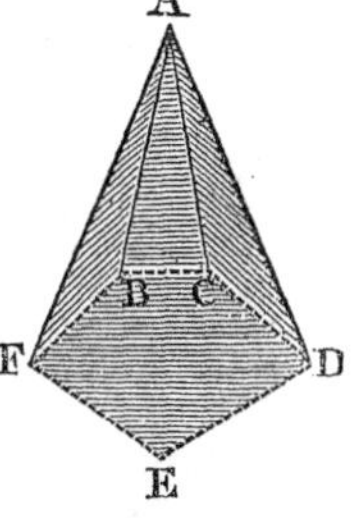

Let A be a solid angle contained by any number of plane angles BAC, CAD, DAE, EAF, FAB; these angles are together less than four right angles.

Let the planes which contain the solid angle at A be cut by another plane, forming the polygon BCDEF. Now, because the solid angle at B is contained by three plane angles, any two of which are greater than the third (Prop. XVII.), the two angles ABC, ABF are greater than the angle FBC. For the same reason, the two angles ACB, ACD are greater than the angle BCD, and so with the other angles of the polygon BCDEF. Hence, the sum of all the angles at the bases of the triangles having the common vertex A, is greater than the sum of all the angles of the polygon BCDEF. But all the angles of these triangles are together equal to twice as many right angles as there are triangles (Prop. XXVII., B. I.), that is, as there are sides of the polygon BCDEF. Also, the angles of the polygon, together with four right angles, are equal to twice as many right angles as the figure has sides (Prop. XXVIII., B. I.); hence all the angles of the triangles are equal to all the angles of the polygon, together with four right angles. But it has been proved that the angles at the bases of the triangles, are greater than the angles of the polygon. Hence the remaining angles of the triangles, viz., those which contain the solid angle at A, are less than four right angles. Therefore, the plane angles, &c.

Scholium. This demonstration supposes that the solid angle is convex; that is, that the plane of neither of the faces, if produced, would cut the solid angle. If it were otherwise, the sum of the plane angles would no longer be limited, and might be of any magnitude.

PROPOSITION XIX. THEOREM.

If two solid angles are contained by three plane angles which are equal, each to each, the planes of the equal angles will be equally inclined to each other.

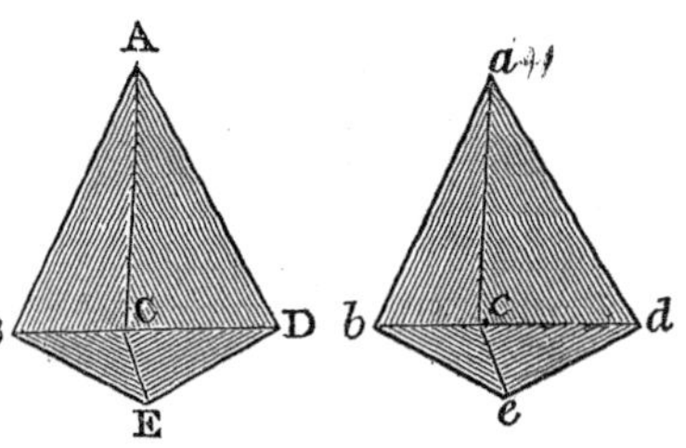

Let A and *a* be two solid angles, contained by three plane angles which are equal, each to each, viz., the angle BAC equal to *bac*, the angle CAD to *cad*, and BAD equal to *bad;* then will the inclination of the planes ABC, ABD be equal to the inclination of the planes *abc*, *abd*.

In the line AC, the common section of the planes ABC, ACD, take any point C; and through C let a plane BCE pass perpendicular to AB, and another plane CDE perpendicular to AD. Also, take *ac* equal to AC; and through *c* let a plane *bce* pass perpendicular to *ab*, and another plane *cde* perpendicular to *ad*.

Now, since the line AB is perpendicular to the plane BCE, it is perpendicular to every straight line which it meets in that plane; hence ABC and ABE are right angles. For the same reason *abc* and *abe* are right angles. Now, in the triangles ABC, *abc*, the angle BAC is, by hypothesis, equal to *bac*, and the angles ABC, *abc* are right angles; therefore the angles ACB, *acb* are equal. But the side AC was made equal to the side *ac;* hence the two triangles are equal (Prop. VII., B. I.); that is, the side AB is equal to *ab*, and BC to *bc*. In the same manner, it may be proved that AD is equal to *ad*, and CD to *cd*.

We can now prove that the quadrilateral ABED is equal to the quadrilateral *abed*. For, let the angle BAD be placed upon the equal angle *bad*, then the point B will fall upon the point *b*, and the point D upon the point *d;* because AB is equal to *ab*, and AD to *ad*. At the same time, BE, which is perpendicular to AB, will fall upon *be*, which is perpendicular to *ab;* and for a similar reason DE will fall upon *de*. Hence the point E will fall upon *e*, and we shall have BE equal to *be*, and DE equal to *de*.

Now, since the plane BCE is perpendicular to the line AB, it is perpendicular to the plane ABD which passes through AB (Prop. VI.). For the same reason CDE is perpendicular to the same plane; hence CE, their common section, is perpendicular to the plane ABD (Prop. VIII.). In the same manner, it may be proved that *ce* is perpendicular to the plane *abd*. Now, in the triangles BCE, *bce*, the angles BEC, *bec* are right angles, the hypothenuse BC is equal to the hypothenuse *bc*, and the side BE is equal to *be;* hence the two triangles are equal, and the angle CBE is equal to the angle *cbe*. But the angle CBE is the inclination of the planes ABC, ABD (Def. 4); and the angle *cbe* is the inclination of the planes *abc*, *abd;* hence these planes are equally inclined to each other.

We must, however, observe that the angle CBE is not, properly speaking, the inclination of the planes ABC, ABD, except when the perpendicular CE falls upon the same side of AB as AD does. If it fall upon the other side of AB, then the angle between the two planes will be obtuse, and this angle, together with the angle B of the triangle CBE, will make two right angles. But in this case, the angle between the two planes *abc*, *abd* will also be obtuse, and this angle, together with the angle *b* of the triangle *cbe*, will also make two right angles. And, since the angle B is always equal to the angle *b*, the inclination of the two planes ABC, ABD will always be equal to that of the planes *abc*, *abd*. Therefore, if two solid angles, &c.

Scholium. If two solid angles are contained by three plane angles which are equal, each to each, and *similarly situated*, the angles will be equal, and will coincide when applied the one to the other. For we have proved that the quadrilateral ABED will coincide with its equal *abed* Now, because the triangle BCE is equal to the triangle *bce*, the line CE, which is perpendicular to the plane ABED, is equal to the line *ce*, which is perpendicular to the plane *abed*. And since only one perpendicular can be drawn to a plane from the same point (Prop. IV., Cor. 2), the lines CE, *ce* must coincide with each other, and the point C coincide with the point *c*. Hence the two solid angles must coincide throughout.

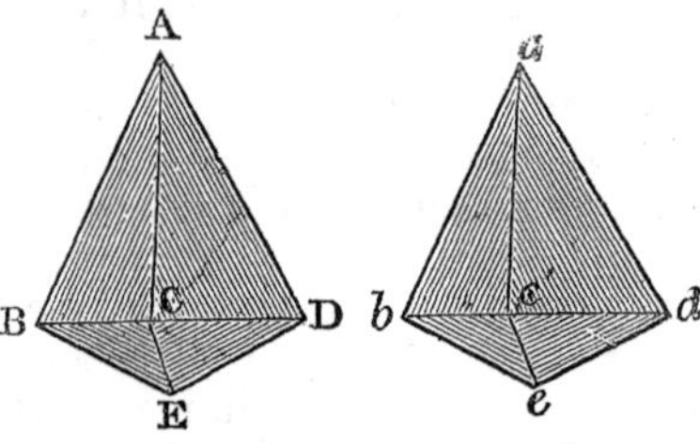

It should, however, be observed that the two solid angles do not admit of superposition, unless the three equal plane angles are *similarly situated* in both cases. For if the perpendiculars CE, *ce* lay on opposite sides of the planes ABED *abed*, the two solid angles could not be made to coincide Nevertheless, the Proposition will always hold true, that the planes containing the equal angles are equally inclined to each other.

BOOK VIII.

POLYEDRONS.

Definitions.

1. A *polyedron* is a solid included by any number of planes which are called its *faces.* If the solid have only four faces, which is the least number possible, it is called a *tetraedron*, if six faces, it is called a *hexaedron;* if eight, an *octaedron*; if twelve, a *dodecaedron;* if twenty, an *icosaedron,* &c.

2. The intersections of the faces of a polyedron are called its *edges.* A *diagonal* of a polyedron is the straight line which joins any two vertices not lying in the same face.

3. *Similar* polyedrons are such as have all their solid angles equal, each to each, and are contained by the same number of similar polygons.

4. A *regular* polyedron is one whose solid angles are all equal to each other, and whose faces are all equal and regular polygons.

5. A *prism* is a polyedron having two faces which are equal and parallel polygons; and the others are parallelograms. The equal and parallel polygons are called the *bases* of the prism; the other faces taken together form the *lateral* or *convex surface.* The *altitude* of a prism is the perpendicular distance between its two bases. The edges which join the corresponding angles of the two polygons are called the *principal edges* of the prism.

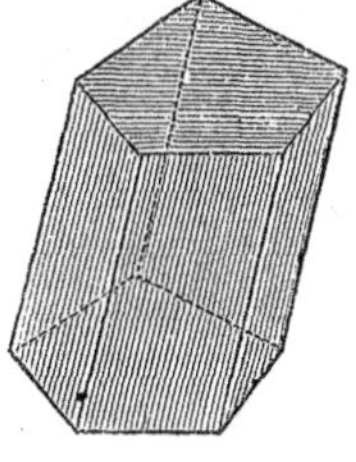

6. A *right prism* is one whose principal edges are all perpendicular to the bases. Any other prism is called an *oblique* prism.

7. A prism is *triangular, quadrangular, pentagonal, hexagonal,* &c., according as its base is a triangle, a quadrilateral, a pentagon, a hexagon, &c.

8. A *parallelopiped* is a prism whose bases are parallelograms. A *right* parallelopiped is one whose faces are all rectangles.

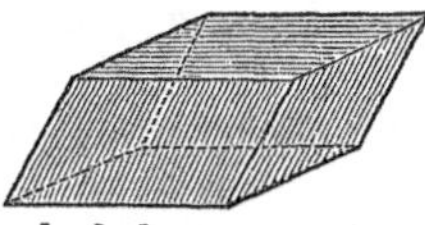

9. A *cube* is a right parallelopiped bounded by six equal squares.

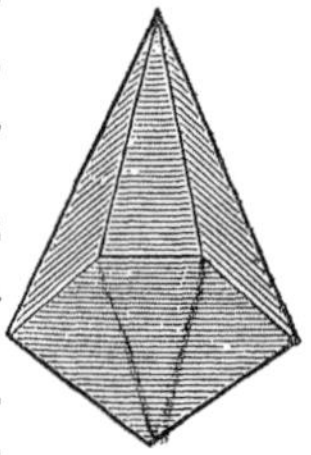

10. A *pyramid* is a polyedron contained by several triangular planes proceeding from the same point, and terminating in the sides of a polygon. This polygon is called the *base* of the pyramid; and the point in which the planes meet, is the *vertex*. The triangular planes form the *convex surface.*

11. The *altitude* of a pyramid is the perpendicular let fall from the vertex upon the plane of the base, produced if necessary. The *slant height* of a pyramid is a line drawn from the vertex, perpendicular to one side of the polygon which forms its base.

12. A pyramid is *triangular*, *quadrangular*, &c., according as the base is a triangle, a quadrilateral, &c.

13. A *regular* pyramid is one whose base is a regular polygon, and the perpendicular let fall from the vertex upon the base, passes through the center of the base. This perpendicular is called the *axis* of the pyramid.

14. A *frustum* of a pyramid is a portion of the solid next the base, cut off by a plane parallel to the base. The *altitude* of the frustum is the perpendicular distance between the two parallel planes.

PROPOSITION I. THEOREM.

The convex surface of a right prism is equal to the perimeter of its base multiplied by its altitude.

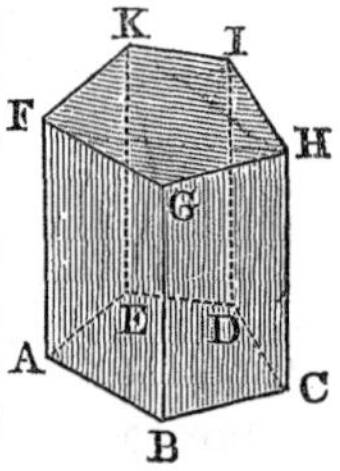

Let ABCDE–K be a right prism; then will its convex surface be equal to the perimeter of the base of AB+BC+CD+DE+EA multiplied by its altitude AF.

For the convex surface of the prism is equal to the sum of the parallelograms AG, BH, CI, &c. Now the area of the parallelogram AG is measured by the product of its base AB by its altitude AF (Prop. IV., Sch., B. IV.). The area of the parallelogram BH is measured by BC×BG; the area of CI is measured by CD×CH, and so of the others. But the lines AF, BG, CH, &c., are all equal to each other (Prop. XIV., B. VII.), and each equal to the altitude of the prism. Also, the lines AB, BC, CD, &c., taken together, from the perimeter of the base of the prism. Therefore, the sum of these parallelograms, or the convex surface of the prism, is equal to the perimeter of its base, multiplied by its altitude.

Cor. If two right prisms have the same altitude, their convex surfaces will be to each other as the perimeters of their bases.

PROPOSITION II. THEOREM.

In every prism, the sections formed by parallel planes are equal polygons.

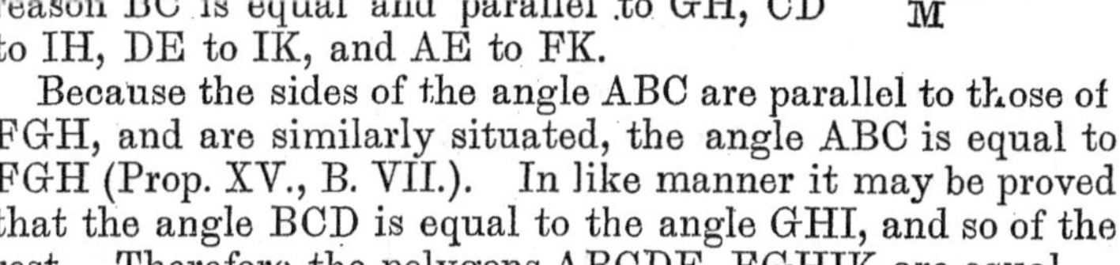

Let the prism LP be cut by the parallel planes AC, FH; then will the sections ABCDE, FGHIK, be equal polygons.

Since AB and FG are the intersections of two parallel planes, with a third plane LMON, they are parallel. The lines AF, BG are also parallel, being edges of the prism; therefore ABGF is a parallelogram, and AB is equal to FG. For the same reason BC is equal and parallel to GH, CD to IH, DE to IK, and AE to FK.

Because the sides of the angle ABC are parallel to those of FGH, and are similarly situated, the angle ABC is equal to FGH (Prop. XV., B. VII.). In like manner it may be proved that the angle BCD is equal to the angle GHI, and so of the rest. Therefore the polygons ABCDE, FGHIK are equal.

Cor. Every section of a prism, made parallel to the base, is equal to the base.

PROPOSITION III. THEOREM.

Two prisms are equal, when they have a solid angle contained by three faces which are equal, each to each, and similarly situated.

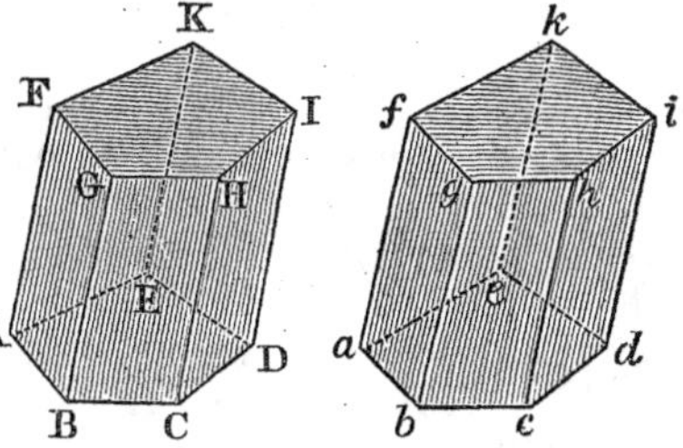

Let AI, *ai* be two prisms having the faces which contain the solid angle B equal to the faces which contain the solid angle *b*; viz., the base ABCDE to the base *abcde*, the parallelogram AG to the parallelogram *ag*, and the parallelogram BH to the parallelogram *bh*; then will the prism AI be equal to the prism *ai*.

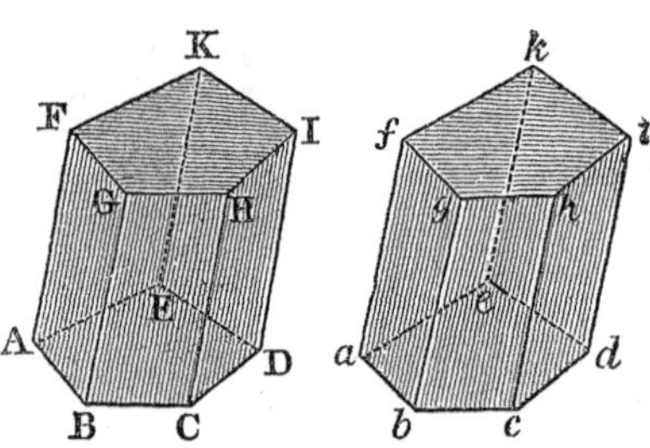

Let the prism AI be applied to the prism *ai*, so that the equal bases AD and *ad* may coincide, the point A falling upon *a*, B upon *b*, and so on. And because the three plane angles which contain the solid angle B, are equal to the three plane angles which contain the solid angle *b*, and these planes are similarly situated, the solid angles B and *b* are equal (Prop. XIX., Sch. B. VII.). Hence the edge BG will coincide with its equal *bg*, and the point G will coincide with the point *g*. Now, because the parallelograms AG and *ag* are equal, the side GF will fall upon its equal *gf*; and for the same reason, GH will fall upon *gh*. Hence the plane of the base FGHIK will coincide with the plane of the base *fghik* (Prop. II., B. VII.). But since the upper bases are equal to their corresponding lower bases, they are equal to each other; therefore the base FI will coincide throughout with *fi*; viz., HI with *hi*, IK with *ik*, and KF with *kf*; hence the prisms coincide throughout, and are equal to each other. Therefore, two prisms, &c.

Cor. Two right prisms, which have equal bases and equal altitudes, are equal.

For, since the side AB is equal to *ab*, and the altitude BG to *bg*, the rectangle ABGF is equal to the rectangle *abgf*. So, also, the rectangle BGHC is equal to the rectangle *bghc*; hence the three faces which contain the solid angle B are equal to the three faces which contain the solid angle *b*; consequently, the two prisms are equal.

PROPOSITION IV. THEOREM.

The opposite faces of a parallelopiped are equal and parallel.

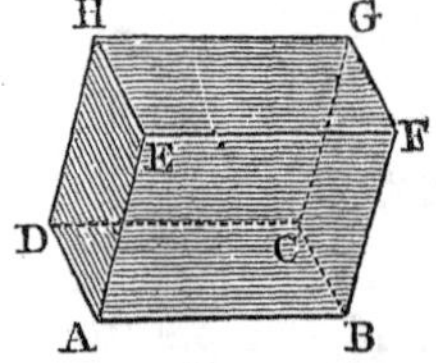

Let ABGH be a parallelopiped; then will its opposite faces be equal and parallel.

From the definition of a parallelopiped (Def. 8) the bases AC, EG are equal and parallel; and it remains to be proved that the same is true of any two opposite faces, as AH, BG. Now, because AC is a parallelogram, the side AD is equal and parallel to BC. For the same reason AE is equal and parallel to BF; hence the angle DAE is equal to the angle CBF

(Prop. XV., B. VII.), and the plane DAE is parallel to the plane CBF. Therefore also the parallelogram AH is equal to the parallelogram BG. In the same manner, it may be proved that the opposite faces AF and DG are equal and parallel. Therefore, the opposite faces, &c.

Cor. 1. Since a parallelopiped is a solid contained by six faces, of which the opposite ones are equal and parallel, any face may be assumed as the base of a parallelopiped.

Cor. 2. *The four diagonals of a parallelopiped bisect each other.*

Draw any two diagonals AG, EC; they will bisect each other. Since AE is equal and parallel to CG, the figure AEGC is a parallelogram; and therefore the diagonals AG, EC bisect each other (Prop. XXXII., B. I.). In the same manner, it may be proved that the two diagonals BH and DF bisect each other; and hence the four diagonals mutually bisect each other, in a point which may be regarded as the center of the parallelopiped.

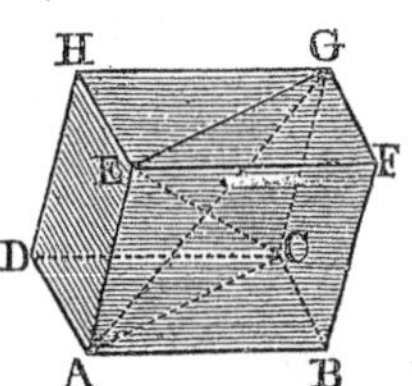

PROPOSITION V. THEOREM.

If a parallelopiped be cut by a plane passing through the diagonals of two opposite faces, it will be divided into two equivalent prisms.

Let AG be a parallelopiped, and AC, EG the diagonals of the opposite parallelograms BD, FH. Now, because AE, CG are each of them parallel to BF, they are parallel to each other; therefore the diagonals AC, EG are in the same plane with AE, CG; and the plane AEGC divides the solid AG into two equivalent prisms.

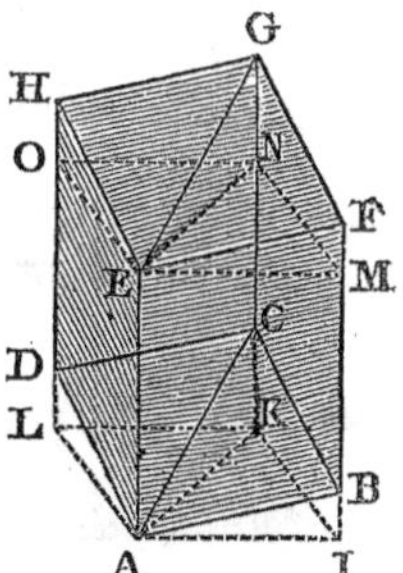

Through the vertices A and E draw the planes AIKL, EMNO perpendicular to AE, meeting the other edges of the parallelopiped in the points I, K, L, and in M, N, O. The sections AIKL, EMNO are equal, because they are formed by planes perpendicular to the same straight line, and, consequently, parallel (Prop. II.). They are also parallelograms, because AI, KL, two opposite sides of the same section, are the intersections of two parallel planes ABFE, DCGH, by the same plane.

For the same reason, the figure ALOE is a parallelogram;

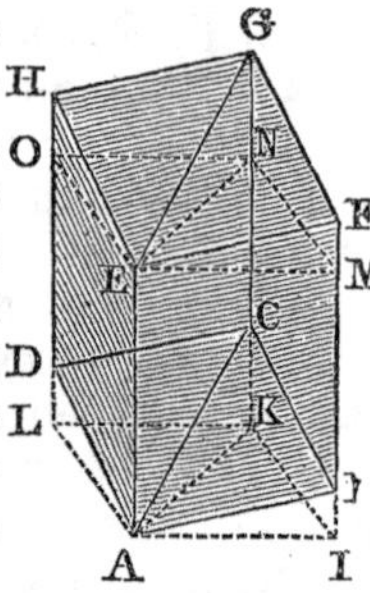

so, also, are AIME, IKNM, KLON, the other lateral faces of the solid AIKL-EMNO; hence this solid is a prism (Def. 5); and it is a right prism because AE is perpendicular to the plane of its base. But the right prism AN is divided into two equal prisms ALK–N, AIK–N; for the basis of these prisms are equal, being halves of the same parallelogram AIKL, and they have the common altitude AE; they are therefore equal (Prop. III. Cor.).

Now, because AEHD, AEOL are parallelograms, the sides DH, LO, being equal to AE, are equal to each other. Take away the common part DO, and we have DL equal to HO. For the same reason, CK is equal to GN. Conceive now that ENO, the base of the solid ENGHO, is placed on AKL, the base of the solid AKCDL; then the point O falling on L and N on K, the lines HO, GN will coincide with their equals DL, CK, because they are perpendiculars to the same plane. Hence the two solids coincide throughout, and are equal to each other. To each of these equals, add the solid ADC–N; then will the oblique prism ADC–G be equivalent to the right prism ALK–N.

In the same manner, it may be proved that the oblique prism ABC–G is equivalent to the right prism AIK–N. But the two right prisms have been proved to be equal; hence the two oblique prisms ADC–G, ABC–G are equivalent to each other. Therefore, if a parallelopiped, &c.

Cor. Every triangular prism is half of a parallelopiped having the same solid angle, and the same edges AB, BC, BF.

Scholium. The triangular prisms into which the oblique parallelopiped is divided, can not be made to coincide, because the plane angles about the corresponding solid angles are not similarly situated.

PROPOSITION VI. THEOREM.

Parallelopipeds, of the same base and the same altitude, are equivalent.

Case first. When their upper bases are between the same parallel lines.

Let the parallelopipeds AG, AL have the base AC common, and let their opposite bases EG, IL be in the same plane, and between the same parallels EK, HL; then will the solid AG be equivalent to the solid AL.

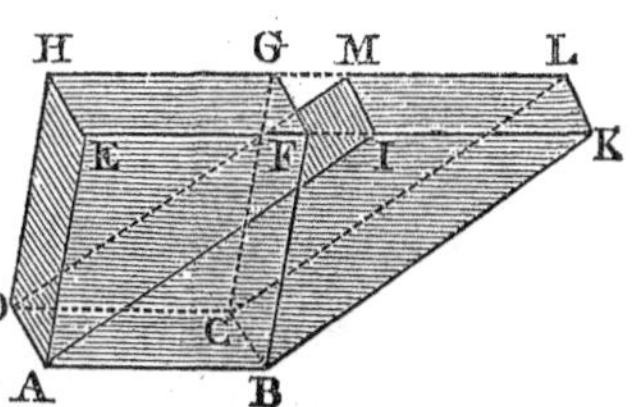

Because AF, AK are parallelograms, EF and IK are each equal to AB, and therefore equal to each other. Hence, if EF and IK be taken away from the same line EK, the remainders EI and FK will be equal. Therefore the triangle AEI is equal to the triangle BFK. Also, the parallelogram EM is equal to the parallelogram FL, and AH to BG. Hence the solid angles at E and F are contained by three faces which are equal to each other and similarly situated; therefore the prism AEI–M is equal to the prism BFK–L (Prop. III.).

Now, if from the whole solid AL, we take the prism AEI–M, there will remain the parallelopiped AL; and if from the same solid AL, we take the prism BFK–L, there will remain the parallelopiped AG. Hence the parallelopipeds AL, AG are equivalent to one another.

Case second. When their upper bases are not between the same parallel lines.

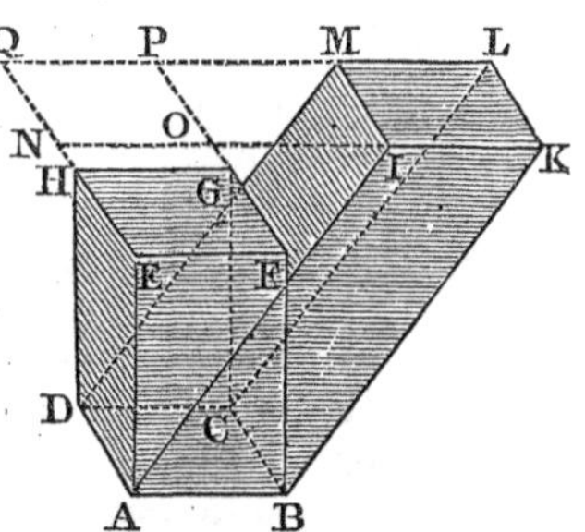

Let the parallelopipeds AG, AL have the same base AC and the same altitude; then will their opposite bases EG, IL be in the same plane. And, since the sides EF and IK are equal and parallel to AB, they are equal and parallel to each other. For the same reason FG is equal and parallel to KL. Produce the sides EH, FG, as also IK, LM, and let them meet in the points N, O, P, Q; the figure NOPQ is a parallelogram equal to each of the bases EG, IL; and, consequently, equal to ABCD, and parallel to it.

Conceive now a third parallelopiped AP, having AC for its lower base, and NP for its upper base. The solid AP will be equivalent to the solid AG, by the first Case, because they have the same lower base, and their upper bases are in the same plane and between the same parallels, EQ, FP. For the same reason, the solid AP is equivalent to the solid AL; hence the solid AG is equivalent to the solid AL. Therefore, parallelopipeds, &c.

PROPOSITION VII. THEOREM.

Any parallelopiped is equivalent to a right parallelopiped having the same altitude and an equivalent base.

Let AL be any parallelopiped; it is equivalent to a right parallelopiped having the same altitude and an equivalent base.

From the points A, B, C, D draw AE, BF, CG, DH, perpendicular to the plane of the lower base, meeting the plane of the upper base in the points E, F, G, H. Join EF, FG, GH, HE; there will thus be formed the parallelopiped AG, equivalent to AL (Prop. VI.); and its lateral faces AF, BG, CH, DE are rectangles. If the base ABCD is also a rectangle, AG will be a right parallelopiped, and it is equivalent to the parallelopiped AL. But if ABCD is not a rectangle, from A and B draw AI, BK perpendicular to CD; and from E and F draw EM, FL perpendicular to GH; and join IM, KL. The solid ABKI–M will be a right parallelopiped. For, by construction, the bases ABKI and EFLM are rectangles; so, also, are the lateral faces, because the edges AE, BF, KL, IM are perpendicular to the plane of the base. Therefore the solid AL is a right parallelopiped. But the two parallelopipeds AG, AL may be regarded as having the same base AF, and the same altitude AI; they are therefore equivalent. But the parallelopiped AG is equivalent to the first supposed parallelopiped; hence this parallelopiped is equivalent to the right parallelopiped AL, having the same altitude, and an equivalent base. Therefore, any parallelopiped, &c.

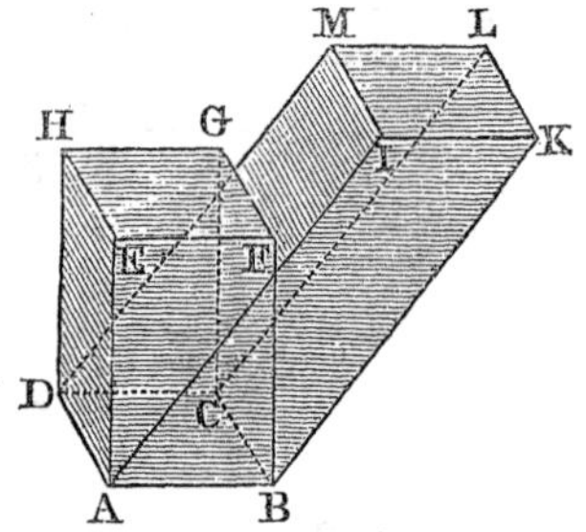

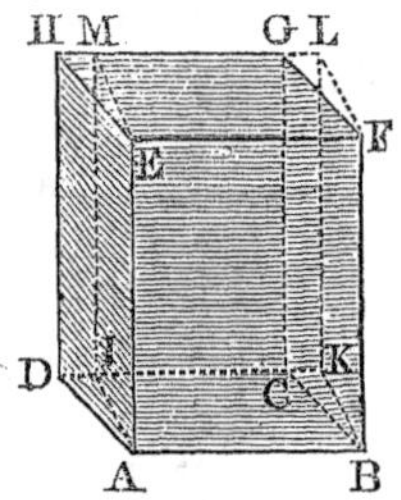

PROPOSITION VIII. THEOREM.

Right parallelopipeds, having the same base, are to each other as their altitudes.

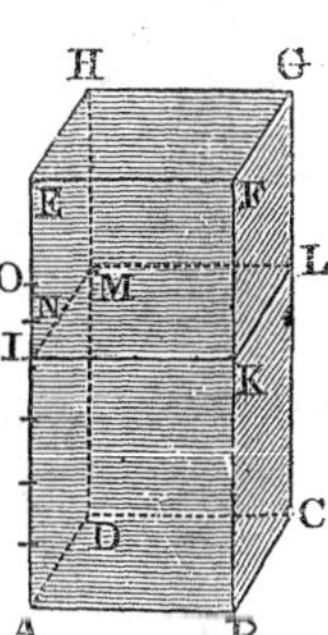

Let AG, AL be two right parallelopipeds having the same base ABCD; then will they be to each other as their altitudes AE, AI.

Case first. When the altitudes are in the ratio of two whole numbers.

Suppose the altitudes AE, AI are in the ratio of two whole numbers; for example, as seven to four. Divide AE into seven equal parts; AI will contain four of those parts. Through the several points of division, let planes be drawn parallel to the base; these planes will divide the solid AG into seven small parallelopipeds, all equal to each other, having equal bases and equal altitudes. The bases are equal, because every section of a prism parallel to the base is equal to the base (Prop. II., Cor.); the altitudes are equal, for these altitudes are the equal divisions of the edge AE. But of these seven equal parallelopipeds, AL contains four; hence the solid AG is to the solid AL, as seven to four, or as the altitude AE is to the altitude AI.

Case second. When the altitudes are not in the ratio of two whole numbers.

Let AG, AL be two parallelopipeds whose altitudes have any ratio whatever; we shall still have the proportion

Solid AG : *solid* AL : : AE : AI.

For if this proportion is not true, the first three terms remaining the same, the fourth term must be greater or less than AI. Suppose it to be greater, and that we have

Solid AG : *solid* AL : : AE : AO.

Divide AE into equal parts each less than OI; there will be at least one point of division between O and I. Designate that point by N. Suppose a parallelopiped to be constructed, having ABCD for its base, and AN for its altitude; and represent this parallelopiped by P. Then, because the altitudes AE, AN are in the ratio of two whole numbers, we shall have, by the preceding Case,

Solid AG : P : : AE : AN.

But, by hypothesis, we have

Solid AG : *solid* AL : : AE : AO.

Hence (Prop. IV., Cor., B. II.),

Solid AL : P : : AO : AN.

But AO is greater than AN; hence the solid AL must be greater than P (Def. 2, B. II.); on the contrary, it is less, which is absurd. Therefore the solid AG can not be to the solid AL, as the line AE to a line greater than AI.

In the same manner, it may be proved that the fourth term of the proportion can not be less than AI; hence it must be AI, and we have the proportion.

Solid AG : *solid* AL : : AE : AI.

Therefore, right parallelopipeds, &c.

PROPOSITION IX. THEOREM.

Right parallelopipeds, having the same altitude, are to each other as their bases.

Let AG, AN be two right parallelopipeds having the same altitude AE; then will they be to each other as their bases; that is,

Solid AG : *solid* AN : : *base* ABCD : *base* AIKL.

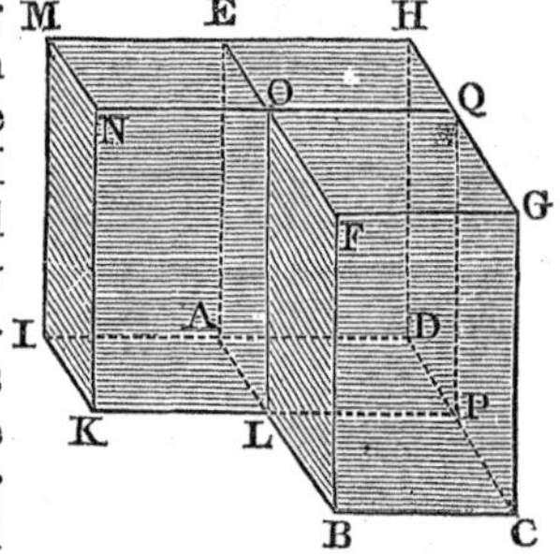

Place the two solids so that their surfaces may have the common angle BAE; produce the plane LKNO till it meets the plane DCGH in the line PQ; a third parallelopiped AQ will thus be formed, which may be compared with each of the parallelopipeds AG, AN. The two solids AG, AQ, having the same base AEHD, are to each other as their altitudes AB, AL (Prop. VIII.); and the two solids AQ, AN, having the same base ALOE, are to each other as their altitudes AD, AI. Hence we have the two proportions

Solid AG : *solid* AQ : : AB : AL;

Solid AQ : *solid* AN : : AD : AI.

Hence (Prop. XI., Cor., B. II.),

Solid AG : *solid* AN : : AB×AD : AL×AI.

But AB×AD is the measure of the base ABCD (Prop. IV., Sch., B. IV.); and AL×AI is the measure of the base AIKL; hence

Solid AG : *solid* AN : : *base* ABCD : *base* AIKL

Therefore, right parallelopipeds, &c.

PROPOSITION X THEOREM.

Any two right parallelopipeds are to each other as the products of their bases by their altitudes.

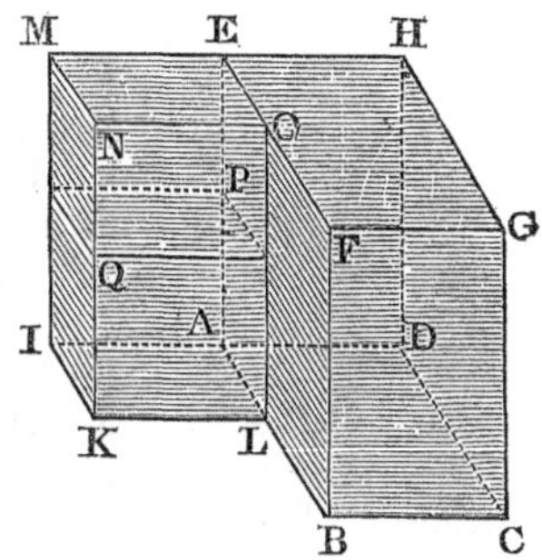

Let AG, AQ be two right parallelopipeds, of which the bases are the rectangles ABCD, AIKL, and the altitudes, the perpendiculars AE, AP; then will the solid AG be to the solid AQ, as the product of ABCD by AE, is to the product of AIKL by AP.

Place the two solids so that their surfaces may have the common angle BAE; produce the planes necessary to form the third parallelopiped AN, having the same base with AQ, and the same altitude with AG. Then, by the last Proposition, we shall have

Solid AG : *solid* AN : : ABCD : AIKL.

But the two parallelopipeds AN, AQ, having the same base AIKL, are to each other as their altitudes AE, AP (Prop. VIII.); hence we have

Solid AN : *solid* AQ : : AE : AP.

Comparing these two proportions (Prop. XI., Cor., B. II.), we have

Solid AG : *solid* AQ : : ABCD×AE : AIKL×AP.

If instead of the base ABCD, we put its equal AB×AD, and instead of AIKL, we put its equal AI×AL, we shall have

Solid AG : *solid* AQ : : AB×AD×AE : AI×AL×AP.

Therefore, any two right parallelopipeds, &c.

Scholium. Hence a right parallelopiped is measured by the product of its base and altitude, or the product of its three dimensions.

It should be remembered, that by the product of two or more lines, we understand the product of the numbers which represent those lines; and these numbers depend upon the linear unit employed, which may be assumed at pleasure. If we take a foot as the unit of measure, then the number of feet in the length of the base, multiplied by the number of feet in its breadth, will give the number of square feet in the base. If we multiply this product by the number of feet in the altitude, it will give the number of cubic feet in the parallelopiped. If we take an inch as the unit of measure, we shall obtain in the same manner the number of cubic inches in the parallelopiped.

PROPOSITION XI. THEOREM.

The solidity of a prism is measured by the product of its base by its altitude.

For any parallelopiped is equivalent to a right parallelopiped, having the same altitude and an equivalent base (Prop. VII.). But the solidity of the latter, is measured by the product of its base by its altitude; therefore the solidity of the former is also measured by the product of its base by its altitude.

Now a triangular prism is half of a parallelopiped having the same altitude and a double base (Prop. V.). But the solidity of the latter is measured by the product of its base by its altitude; hence a triangular prism is measured by the product of its base by its altitude.

But any prism can be divided into as many triangular prisms of the same altitude, as there are triangles in the polygon which forms its base. Also, the solidity of each of these triangular prisms, is measured by the product of its base by its altitude; and since they all have the same altitude, the sum of these prisms will be measured by the sum of the triangles which form the bases, multiplied by the common altitude. Therefore, the solidity of any prism is measured by the product of its base by its altitude.

Cor. If two prisms have the same altitude, the products of the bases by the altitudes, will be as the bases (Prop. VIII., B. II.); hence *prisms of the same altitude are to each other as their bases.* For the same reason, *prisms of the same base are to each other as their altitudes;* and *prisms generally are to each other as the products of their bases and altitudes.*

PROPOSITION XII. THEOREM.

Similar prisms are to each other as the cubes of their homologous edges.

Let ABCDE–F, *abcde–f* be two similar prisms; then will the prism AD–F be to the prism *ad–f*, as AB^3 to ab^3, or as AF^3 to af^3.

For the solids are to each other as the products of their bases and altitudes (Prop. XI., Cor.); that is, as $ABCDE \times AF$, to $abcde \times af$. But since the prisms are similar, the bases are similar figures, and are to each other as the squares of

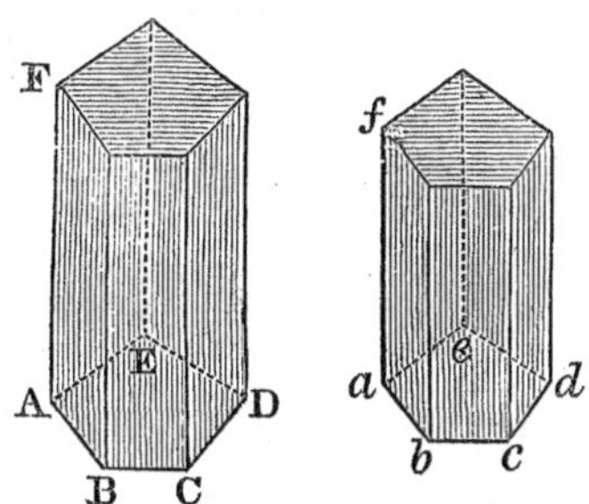

their homologous sides; that is, as AB^2 to ab^2. Therefore, we have

$$Solid\ FD : solid\ fd :: AB^2 \times AF : ab^2 \times af.$$

But since BF and *bf* are similar figures, their homologous sides are proportional; that is,

$$AB : ab :: AF : af,$$

whence (Prop. X., B. II.),

$$AB^2 : ab^2 :: AF^2 : af^2.$$

Also $$AF : af :: AF : af.$$

Therefore (Prop. XI., B. II.),

$$AB^2 \times AF : ab^2 \times af :: AF^3 : af^3 :: AB^3 : ab^3.$$

Hence (Prop. IV., B. II.), we have

$$Solid\ FD : solid\ fd :: AB^3 : ab^3 :: AF^3 : af^3.$$

Therefore, similar prisms, &c.

PROPOSITION XIII. THEOREM.

If a pyramid be cut by a plane parallel to its base,

1*st. The edges and the altitude will be divided proportionally.*

2*d. The section will be a polygon similar to the base.*

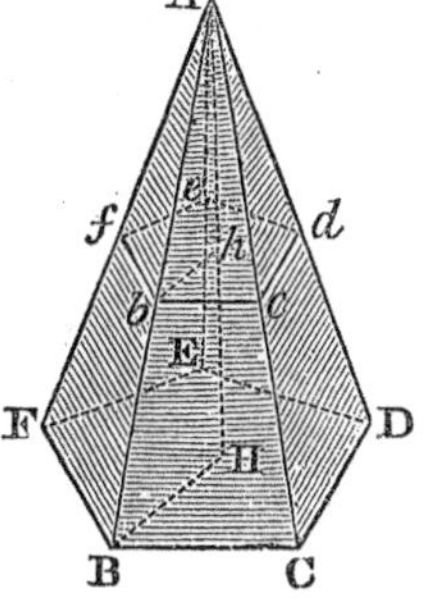

Let A–BCDEF be a pyramid cut by a plane *bcdef* parallel to its base, and let AH be its altitude; then will the edges AB, AC, AD, &c., with the altitude AH, be divided proportionally in *b, c, d, e, f, h;* and the section *bcdef* will be similar to BCDEF.

First. Since the planes FBC, *fbc* are parallel, their sections FB, *fb* with a third plane AFB are parallel (Prop. XII., B. VII.); therefore the triangles AFB, A*fb* are similar, and we have the proportion

$$AF : Af :: AB : Ab.$$

For the same reason,

$$AB : Ab :: AC : Ac,$$

and so for the other edges. Therefore the edges AB, AC, &c., are cut proportionally in b, c, &c. Also, since BH and bh are parallel, we have

$$AH : Ah :: AB : Ab.$$

Secondly Because fb is parallel to FB, bc to BC, cd to CD &c., the angle fbc is equal to FBC (Prop. XV., B. VII.), the angle bcd is equal to BCD, and so on. Moreover, since the triangles AFB, Afb are similar, we have

$$FB : fb :: AB \cdot Ab.$$

And because the triangles ABC, Abc are similar, we have

$$AB : Ab :: BC : bc.$$

Therefore, by equality of ratios (Prop. IV., B. II.),

$$FB : fb :: BC : bc.$$

For the same reason,

$$BC : bc :: CD : cd, \text{ and so on.}$$

Therefore the polygons BCDEF, $bcdef$ have their angles equal, each to each, and their homologous sides proportional; hence they are similar. Therefore, if a pyramid, &c.

Cor. 1. *If two pyramids, having the same altitude, and their bases situated in the same plane, are cut by a plane parallel to their bases, the sections will be to each other as the bases.*

Let A–BCDEF, A–MNO be two pyramids having the same altitude, and their bases situated in the same plane; if these pyramids are cut by a plane parallel to the bases, the sections $bcdef$, mno will be to each other as the bases BCDEF, MNO.

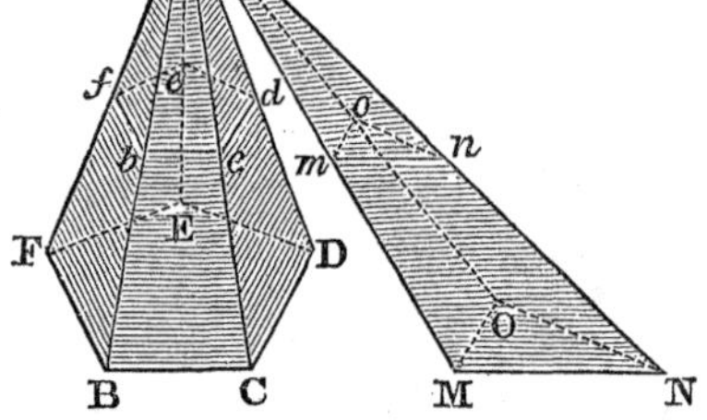

For, since the polygons BCDEF, $bcdef$ are similar, their surfaces are as the squares of the homologous sides BC bc (Prop. XXVI., B. IV.). But, by the preceding Proposition

$$BC : bc :: AB : Ab.$$

Therefore, $\quad BCDEF : bcdef :: AB^2 : Ab^2.$

For the same reason,

$$MNO : mno :: AM^2 : Am^2.$$

But since $bcdef$ and mno are in the same plane, we have

$$AB : Ab :: AM : Am \text{ (Prop. XVI., B. VII.)};$$

consequently, $\quad BCDEF : bcdef :: MNO : mno.$

Cor. 2. If the bases BCDEF, MNO are equivalent, the sections $bcdef$, mno will also be equivalent.

PROPOSITION XIV. THEOREM.

The convex surface of a regular pyramid, is equal to the perimeter of its base, multiplied by half the slant height.

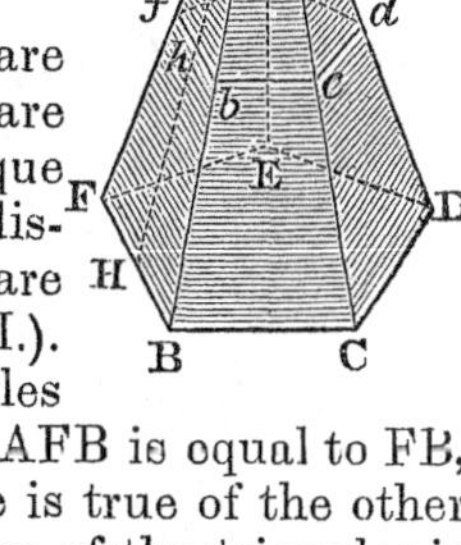

Let A–BDE be a regular pyramid, whose base is the polygon BCDEF, and its slant height AH; then will its convex surface be equal to the perimeter BC+CD+DE, &c., multiplied by half of AH.

The triangles AFB, ABC, ACD, &c., are all equal for the sides FB, BC, CD, &c., are all equal, (Def. 13); and since the oblique lines AF, AB, AC, &c., are all at equal distances from the perpendicular, they are equal to each other (Prop. V., B. VII.). Hence the altitudes of these several triangles are equal. But the area of the triangle AFB is equal to FB, multiplied by half of AH; and the same is true of the other triangles ABC, ACD, &c. Hence the sum of the triangles is equal to the sum of the bases FB, BC, CD, DE, EF, multiplied by half the common altitude AH; that is, the convex surface of the pyramid is equal to the perimeter of its base, multiplied by half the slant height.

Cor. 1. *The convex surface of a frustum of a regular pyramid is equal to the sum of the perimeters of its two bases, multiplied by half its slant height.*

Each side of a frustum of a regular pyramid, as FB*bf*, is a trapezoid (Prop. XIII.). Now the area of this trapezoid is equal to the sum of its parallel sides FB, *fb*, multiplied by half its altitude H*h* (Prop. VII., B. IV.). But the altitude of each of these trapezoids is the same; therefore the area of all the trapezoids, or the convex surface of the frustum, is equal to the sum of the perimeters of the two bases, multiplied by half the slant height.

Cor. 2. If the frustum is cut by a plane, parallel to the bases, and at equal distances from them, this plane must bisect the edges B*b*, C*c*, &c. (Prop. XVI., B. IV.); and the area of each trapezoid is equal to its altitude, multiplied by the line which joins the middle points of its two inclined sides (Prop. VII., Cor., B. IV.). Hence *the convex surface of a frustum of a pyramid is equal to its slant height, multiplied by the perimeter of a section at equal distances between the two bases.*

PROPOSITION XV. THEOREM.

Triangular pyramids, having equivalent bases and equal altitudes, are equivalent.

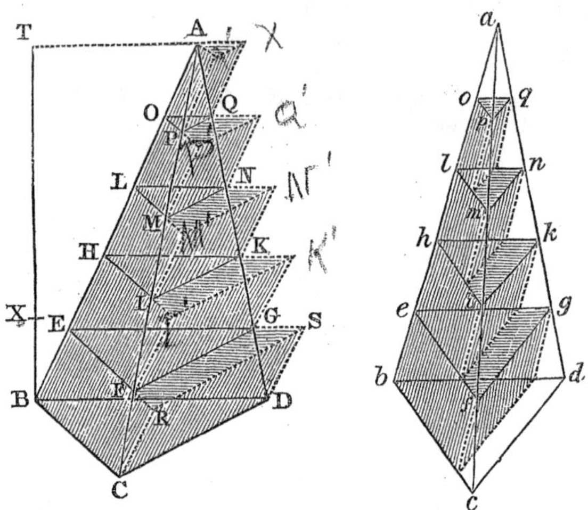

Let A–BCD, *a–bcd* be two triangular pyramids having equivalent bases BCD, *bcd*, supposed to be situated in the same plane, and having the common altitude TB; then will the pyramid A–BCD be equivalent to the pyramid *a–bcd*.

For, if they are not equivalent, let the pyramid A–BCD exceed the pyramid *a–bcd* by a prism whose base is BCD and altitude BX.

Divide the altitude BT into equal parts, each less than BX; and through the several points of division, let planes be made to pass parallel to the base BCD, making the sections EFG, *efg* equivalent to each other (Prop. XIII., Cor. 2): also, HIK equivalent to *hik*, &c.

From the point C, draw the straight line CR parallel to BE, meeting EF produced in R; and from D draw DS parallel to BE, meeting EG in S. Join RS, and it is plain that the solid BCD–ERS is a prism lying partly without the pyramid. In the same manner, upon the triangles EFG, HIK, &c., taken as bases, construct exterior prisms, having for edges the parts EH, HL, &c., of the line AB. In like manner, on the bases *efg*, *hik*, *lmn*, &c., in the second pyramid, construct interior prisms, having for edges the corresponding parts of *ab*. It is plain that the sum of all the exterior prisms of the pyramid A–BCD is greater than this pyramid; and, also, that the sum of all the interior prisms of the pyramid *a–bcd* is smaller than this pyramid. Hence the difference between the sum of all the exterior prisms, and the sum of all the interior ones, must be greater than the difference between the two pyramids themselves.

Now, beginning with the bases BCD, *bcd*, the second exterior prism EFG–H is equivalent to the first interior prism *efg–b*, because their bases are equivalent, and they have the same altitude. For the same reason, the third exterior prism HIK–L and the second interior prism *hik–e* are equivalent; the fourth exterior and the third interior; and so on, to the last in each series. Hence all the exterior prisms of the pyramid A–BCD, excepting the first prism BCD–E, have equivalent corresponding ones in the interior prisms of the pyramid *a–bcd*. Therefore the prism BCD–E is the difference between the sum of all the exterior prisms of the pyramid A–BCD, and the sum of all the interior prisms of the pyramid *a–bcd*. But the difference between these two sets of prisms has been proved to be greater than that of the two pyramids; hence the prism BCD–E is greater than the prism BCD–X; which is impossible, for they have the same base BCD, and the altitude of the first, is less than BX, the altitude of the second. Hence the pyramids A–BCD, *a–bcd* are not unequal; that is, they are equivalent to each other. Therefore, triangular pyramids, &c.

PROPOSITION XVI. THEOREM.

Every triangular pyramid is the third part of a triangular prism having the same base and the same altitude.

Let E–ABC be a triangular pyramid, and ABC–DEF a triangular prism having the same base and the same altitude; then will the pyramid be one third of the prism.

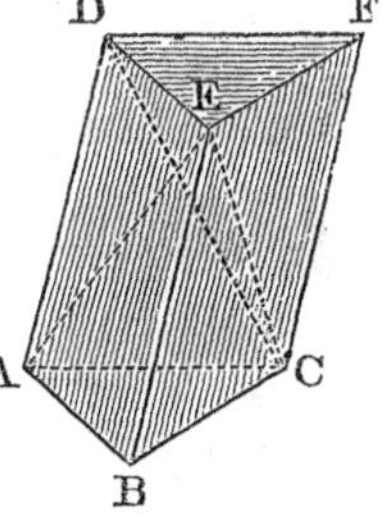

Cut off from the prism the pyramid E–ABC by the plane EAC; there will remain the solid E–ACFD, which may be considered as a quadrangular pyramid whose vertex is E, and whose base is the parallelogram ACFD. Draw the diagonal CD, and through the points C, D, E pass a plane, dividing the quadrangular pyramid into two triangular ones E–ACD E–CFD. Then, because ACFD is a parallelogram, of which

CD is the diagonal, the triangle ACD is equal to the triangle CDF. Therefore the pyramid, whose base is the triangle ACD, and vertex the point E, is equivalent to the pyramid whose base is the triangle CDF, and vertex the point E. But the latter pyramid is equivalent to the pyramid E–ABC for they have equal bases, viz., the triangles ABC, DEF, and the same altitude, viz., the altitude of the prism ABC–DEF. Therefore the three pyramids E–ABC, E–ACD, E–CDF, are equivalent to each other, and they compose the whole prism ABC–DEF; hence the pyramid E–ABC is the third part of the prism which has the same base and the same altitude.

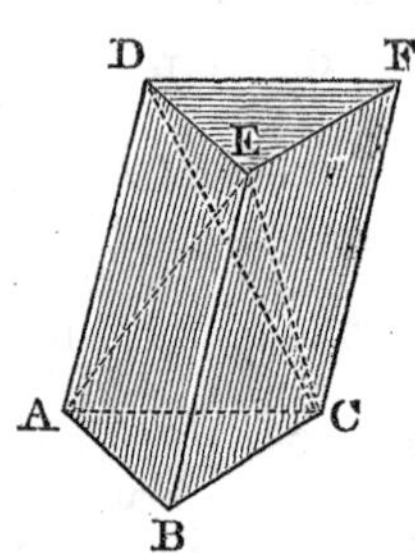

Cor. The solidity of a triangular pyramid is measured by the product of its base by one third of its altitude.

PROPOSITION XVII. THEOREM.

The solidity of every pyramid is measured by the product of its base by one third of its altitude.

Let A–BCDEF be any pyramid, whose base is the polygon BCDEF, and altitude AH; then will the solidity of the pyramid be measured by BCDEF $\times \frac{1}{3}$AH.

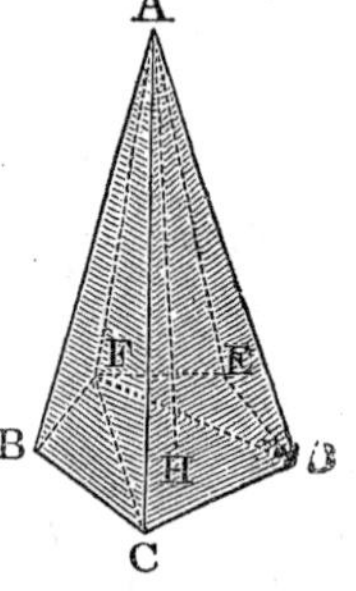

Divide the polygon BCDEF into triangles by the diagonals CF, DF; and let planes pass through these lines and the vertex A; they will divide the polygonal pyramid A–BCDEF into triangular pyramids, all having the same altitude AH. But each of these pyramids is measured by the product of its base by one third of its altitude (Prop. XVI., Cor.); hence the sum of the triangular pyramids, or the polygonal pyramid A–BCDEF, will be measured by the sum of the triangles BCF, CDF, DEF, or the polygon BCDEF, multiplied by one third of AH. Therefore every pyramid is measured by the product of its base by one third of its altitude.

Cor. 1. Every pyramid is one third of a prism having the same base and altitude.

Cor. 2. Pyramids of the same altitude are to each other as their bases; pyramids of the same base are to each other as their altitudes; and pyramids generally are to each other as the products of their bases by their altitudes.

Cor. 3. Similar pyramids are to each other as the cubes of their homologous edges.

Scholium. The solidity of any polyedron may be found by dividing it into pyramids, by planes passing through its vertices.

PROPOSITION XVIII. THEOREM.

A frustum of a pyramid is equivalent to the sum of three pyramids, having the same altitude as the frustum, and whose bases are the lower base of the frustum, its upper base, and a mean proportional between them.

Case first. When the base of the frustum is a triangle.

Let ABC–DEF be a frustum of a triangular pyramid. If a plane be made to pass through the points A, C, E, it will cut off the pyramid E–ABC, whose altitude is the altitude of the frustum, and its base is ABC, the lower base of the frustum.

Pass another plane through the points C, D, E; it will cut off the pyramid C–DEF, whose altitude is that of the frustum, and its base is DEF, the upper base of the frustum.

To find the magnitude of the remaining pyramid E–ACD, draw EG parallel to AD; join CG, DG. Then, because the two triangles AGC, DEF have the angles at A and D equal to each other, we have (Prop. XXIII., B. IV.)

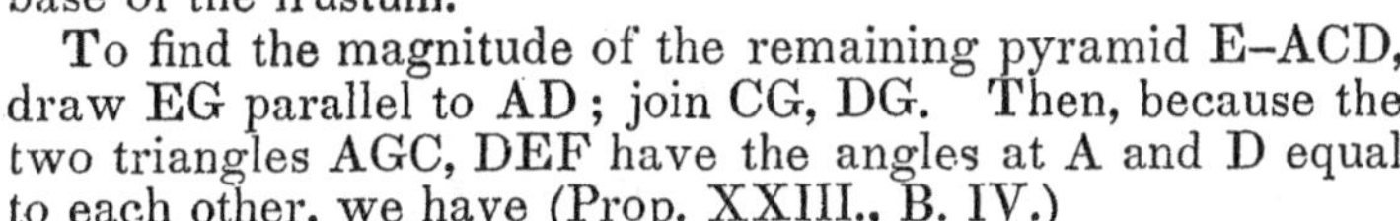

AGC : DEF : : AG × AC : DE × DF,

: : AC : DF, because AG is equal to DE.

Also (Prop. VI., Cor. 1, B. IV.),

ACB : ACG : : AB : AG or DE.

But, because the triangles ABC, DEF are similar (Prop. XIII.), we have

AB : DE : : AC : DF.

Therefore (Prop. IV., B. II.),

ACB : ACG : : ACG : DEF;

that is, the triangle ACG is a mean proportional between ACB and DEF, the two bases of the frustum.

Now the pyramid E–ACD is equivalent to the pyramid G–ACD, because it has the same base and the same altitude; for EG is parallel to AD, and, consequently, parallel to the plane ACD. But the pyramid G–ACD has the same altitude as the frustum, and its base ACG is a mean proportional between the two bases of the frustum.

Case second. When the base of the frustum is any polygon.

Let BCDEF–*bcdef* be a frustum of any pyramid.

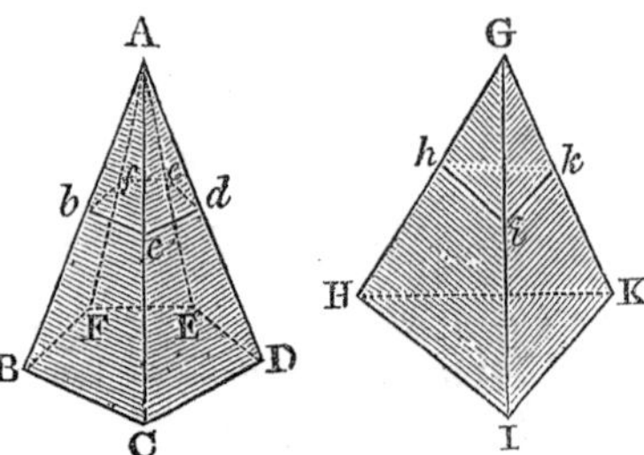

Let G–HIK be a triangular pyramid having the same altitude and an equivalent base with the pyramid A–BCDEF, and from it let a frustum HIK–*hik* be cut off, having the same altitude with the frustum BCDEF–*bcdef*. The entire pyramids are equivalent (Prop. XVII.) and the small pyramids A–*bcdef*, G–*hik* are also equivalent, for their altitudes are equal, and their bases are equivalent (Prop. XIII., Cor. 2). Hence the two frustums are equivalent, and they have the same altitude, with equivalent bases. But the frustum HIK–*hik* has been proved to be equivalent to the sum of three pyramids, each having the same altitude as the frustum, and whose bases are the lower base of the frustum, its upper base, and a mean proportional between them. Hence the same must be true of the frustum of any pyramid Therefore, a frustum of a pyramid, &c.

PROPOSITION XIX. THEOREM.

There can be but five regular polyedrons.

Since the faces of a regular polyedron are regular polygons, they must consist of equilateral triangles, of squares, of regular pentagons, or polygons of a greater number of sides.

First. If the faces are equilateral triangles, each solid angle of the polyedron may be contained by three of these tri

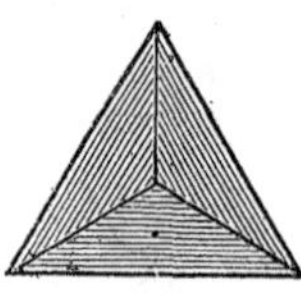
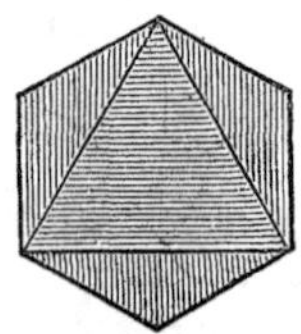
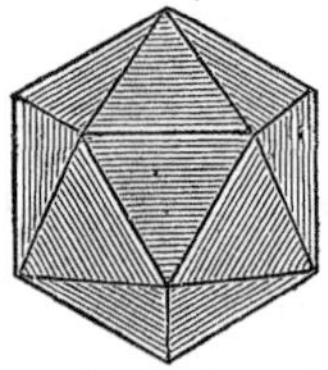

angles, forming the *tetraedron;* or by four, forming the *octaedron;* or by five, forming the *icosaedron.*

No other regular polyedron can be formed with equilateral triangles; for six angles of these triangles amount to four right angles, and can not form a solid angle (Prop. XVIII., B. VII.).

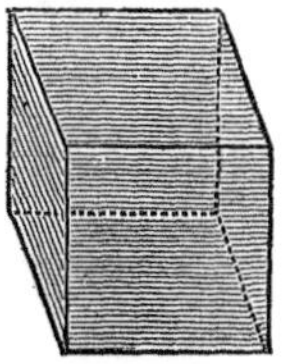

Secondly. If the faces are squares, their angles may be united three and three, forming the *hexaedron*, or cube.

Four angles of squares amount to four right angles, and can not form a solid angle.

Thirdly. If the faces are regular pentagons, their angles may be united three and three, forming the regular *dodecaedron.* Four angles of a regular pentagon, are greater than four right angles, and can not form a solid angle.

Fourthly. A regular polyedron can not be formed with regular hexagons, for three angles of a regular hexagon amount to four right angles. Three angles of a regular heptagon amount to more than four right angles; and the same is true of any polygon having a greater number of sides.

Hence there can be but five regular polyedrons; three formed with equilateral triangles, one with squares, and one with pentagons

BOOK IX.

SPHERICAL GEOMETRY

Definitions.

1. A *sphere* is a solid bounded by a curved surface, all the points of which are equally distant from a point within, called the *center*.

The sphere may be conceived to be described by the revolution of a semicircle ADB, about its diameter AB, which remains unmoved.

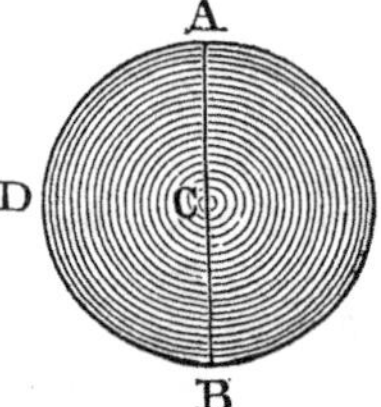

2. The *radius* of a sphere, is a straight line drawn from the center to any point of the surface. The *diameter*, or *axis*, is a line passing through the center, and terminated each way by the surface.

All the radii of a sphere are equal; all the diameters are also equal, and each double of the radius.

3. It will be shown (Prop. I.), that every section of a sphere made by a plane is a circle. A *great circle* is a section made by a plane which passes through the center of the sphere. Any other section made by a plane is called a *small circle.*

4. A plane *touches* a sphere, when it meets the sphere, but, being produced, does not cut it.

5. The *pole* of a circle of a sphere, is a point in the surface equally distant from every point in the circumference of this circle. It will be shown (Prop. V.), that every circle, whether great or small, has two poles.

6. A *spherical triangle* is a part of the surface of a sphere, bounded by three arcs of great circles, each of which is less than a semicircumference. These arcs are called the *sides* of the triangle; and the angles which their planes make with each other, are the angles of the triangle.

7. A spherical triangle is called *right-angled*, *isosceles* or *equilateral*, in the same cases as a plane triangle.

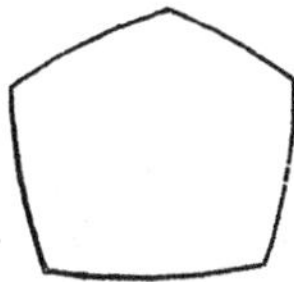

8. A *spherical polygon* is a part of the surface of a sphere bounded by several arcs of great circles.

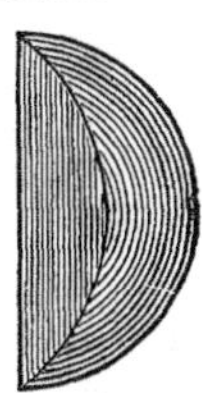

9. A *lune* is a part of the surface of a sphere included between the halves of two great circles.

10. A *spherical wedge*, or *ungula*, is that portion of the sphere included between the same semicircles, and has the lune for its base.

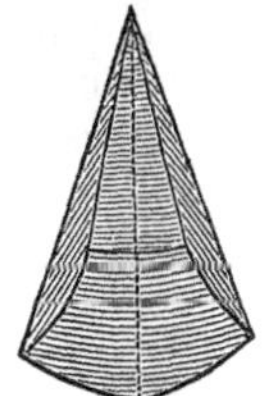

11. A *spherical pyramid* is a portion of the sphere included between the planes of a solid angle, whose vertex is at the center. The *base* of the pyramid is the spherical polygon intercepted by those planes.

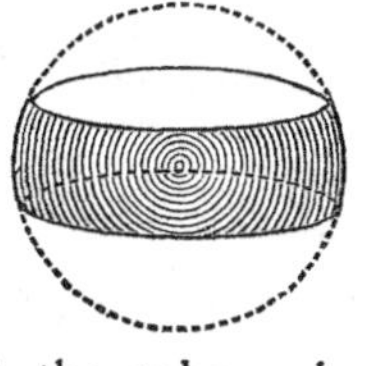

12. A *zone* is a part of the surface of a sphere included between two parallel planes.

13. A *spherical segment* is a portion of the sphere included between two parallel planes.

14. The *bases* of the segment are the sections of the sphere; the *altitude* of the segment, or zone, is the distance between the sections. One of the two planes may *touch* the sphere, in which case the segment has but one base.

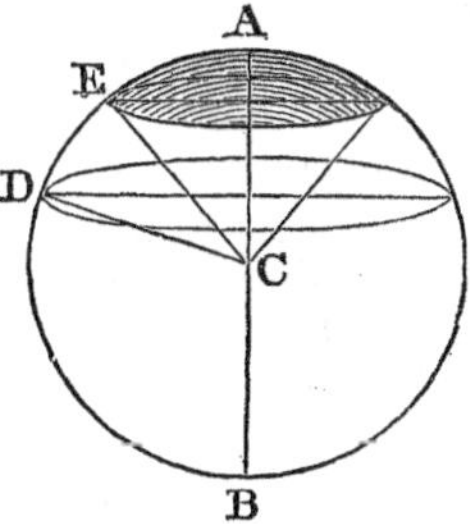

15. A *spherical sector* is a solid described by the revolution of a circular sector, in the same manner as the sphere is described by the revolution of a semicircle.

While the semicircle ADB, revolving round its diameter AB, describes a sphere, every circular sector, as ACE or ECD, describes a spherical sector.

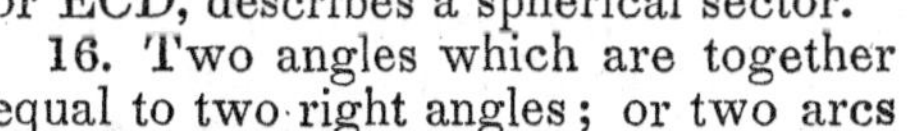

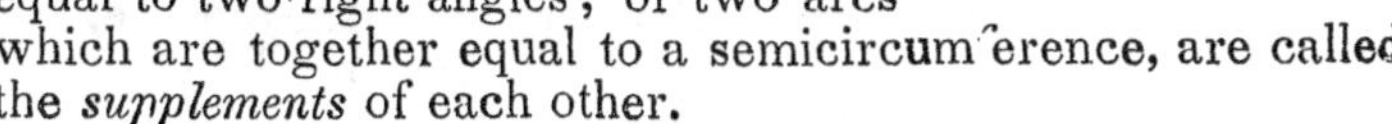

16. Two angles which are together equal to two right angles; or two arcs which are together equal to a semicircumference, are called the *supplements* of each other.

PROPOSITION I. THEOREM.

Every section of a sphere, made by a plane, is a circle

Let ABD be a section, made by a plane, in a sphere whose center is C. From the point C draw CE perpendicular to the plane ABD; and draw lines CA, CB, CD, &c., to different points of the curve ABD which bounds the section.

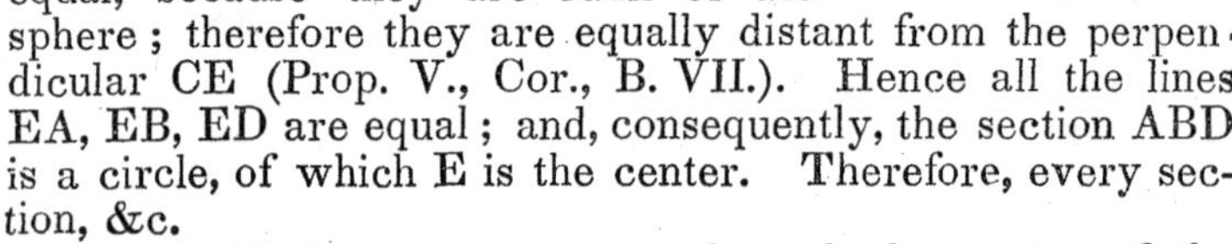

The oblique lines CA, CB, CD are equal, because they are radii of the sphere; therefore they are equally distant from the perpendicular CE (Prop. V., Cor., B. VII.). Hence all the lines EA, EB, ED are equal; and, consequently, the section ABD is a circle, of which E is the center. Therefore, every section, &c.

Cor. 1. If the section passes through the center of the sphere, its radius will be the radius of the sphere; hence all great circles of a sphere are equal to each other.

Cor. 2. Two great circles always bisect each other; for, since they have the same center, their common section is a diameter of both, and therefore bisects both.

Cor. 3. Every great circle divides the sphere and its surface into two equal parts. For if the two parts are separated and applied to each other, base to base, with their convexities turned the same way, the two surfaces must coincide; otherwise there would be points in these surfaces unequally distant from the center.

Cor. 4. The center of a small circle, and that of the sphere, are in a straight line perpendicular to the plane of the small circle.

Cor. 5. The circle which is furthest from the center is the least; for the greater the distance CE, the less is the chord AB, which is the diameter of the small circle ABD.

Cor. 6. An arc of a great circle may be made to pass through any two points on the surface of a sphere; for the two given points, together with the center of the sphere, make three points which are necessary to determine the position of a plane. If, however, the two given points were situated at the extremities of a diameter, these two points and the center would then be in one straight line, and any number of great circles might be made to pass through them.

PROPOSITION II. THEOREM.

Any two sides of a spherical triangle are greater than the third.

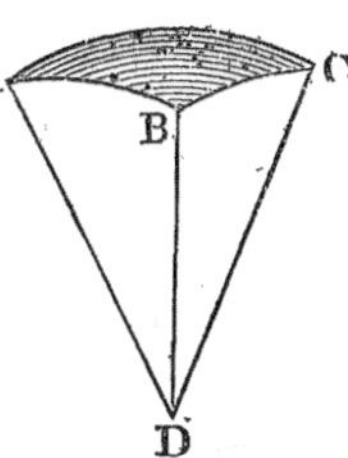

Let ABC be a spherical triangle; any two sides as, AB, BC, are together greater than the third side AC.

Let D be the center of the sphere; and join AD, BD, CD. Conceive the planes ADB, BDC, CDA to be drawn, forming a solid angle at D. The angles ADB, BDC, CDA will be measured by AB, BC, CA, the sides of the spherical triangle. But when a solid angle is formed by three plane angles, the sum of any two of them is greater than the third (Prop. XVII., B. VII.); hence any two of the arcs AB, BC, CA must b greater than the third. Therefore, any two sides, &c.

PROPOSITION III. THEOREM.

The shortest path from one point to another on the surface of a sphere, is the arc of a great circle joining the two given points.

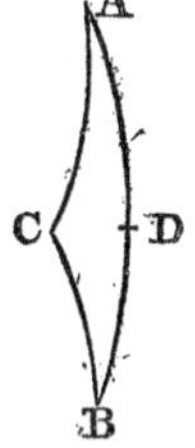

Let A and B be any two points on the surface of a sphere, and let ADB be the arc of a great circle which joins them; then will the line ADB be the shortest path from A to B on the surface of the sphere.

For, if possible, let the shortest path from A to B pass through C, a point situated out of the arc of a great circle ADB. Draw AC, CB, arcs of great circles, and take BD equal to BC.

By the preceding theorem, the arc ADB is less than AC+CB. Subtracting the equal arcs BD and BC, there will remain AD less than AC. Now the shortest path from B to C, whether it be an arc of a great circle, or some other line, is equal to the shortest path from B to D; for, by revolving BC around B, the point C may be made to coincide with D, and thus the shortest path from B to C must coincide with the shortest path from B to D. But the shortest path from A to B was supposed to pass through C; hence the shortest path from A to C, can not be greater than the shortest path from A to D.

Now the arc AD has been proved to be less than AC; and therefore if AC be revolved about A until the point C falls on the arc ADB, the point C will fall between D and B. Hence the shortest path from C to A must be greater than the shortest path from D to A; but it has just been proved not to be greater, which is absurd. Consequently, no point of the shortest path from A to B, can be out of the arc of a great circle ADB. Therefore, the shortest path, &c.

PROPOSITION IV. THEOREM.

The sum of the sides of a spherical polygon, is less than the circumference of a great circle.

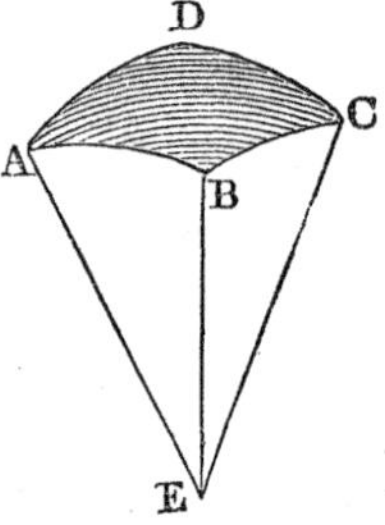

Let ABCD be any spherical polygon; then will the sum of the sides AB, BC, CD, DA be less than the circumference of a great circle.

Let E be the center of the sphere, and join AE, BE, CE, DE. The solid angle at E is contained by the plane angles AEB, BEC, CED, DEA, which together are less than four right angles (Prop. XVIII., B. VII.). Hence the sides AB, BC, CD, DA, which are the measures of these angles, are together less than four quadrants described with the radius AE; that is, than the circumference of a great circle. Therefore, the sum of the sides, &c.

PROPOSITION V. THEOREM.

The extremities of a diameter of a sphere, are the poles of all circles perpendicular to that diameter.

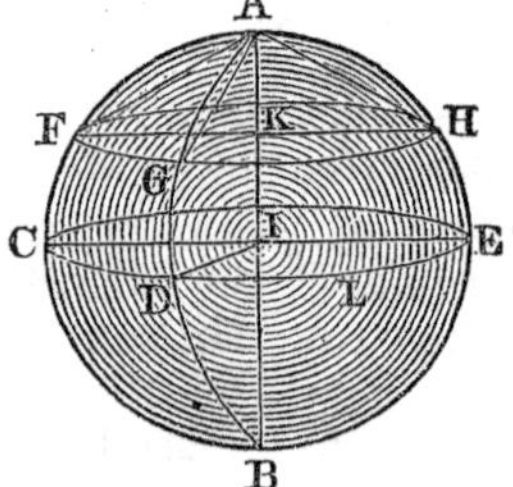

Let AB be a diameter perpendicular to CDE, a great circle of a sphere, and also to the small circle FGH; then will A and B, the extremities of the diameter, be the poles of both these circles.

For, because AB is perpendicular to the plane CDE, it is perpendicular to every straight line CI, DI, EI, &c., drawn through its foot in the plane; hence all the arcs AC, AD, AE, &c., are quarters of the cir cumference. So, also, the arcs BC, BD, BE, &c., are quarters of the circumference; hence the points A and B are each equally distant from all the points of the circumference CDE; they are, therefore, the poles of that circumference (Def. 5).

Secondly. Because the radius AI is perpendicular to the plane of the circle FGH, it passes through K, the center of that circle (Prop. I., Cor. 4). Hence, if we draw the oblique lines AF, AG, AH, these lines will be equally distant from the perpendicular AK, and will be equal to each other (Prop. V., B. VII.). But since the chords AF, AG, AH are equal, the arcs are equal; hence the point A is a pole of the small circle FGH; and in the same manner it may be proved that B is the other pole.

Cor. 1. The arc of a great circle AD, drawn from the pole to the circumference of another great circle CDE, is a quadrant; and this quadrant is perpendicular to the arc CD. For, because AI is perpendicular to the plane CDI, every plane ADB which passes through the line AI is perpendicular to the plane CDI (Prop. VI., B. VII.); therefore the angle contained by these planes, or the angle ADC (Def. 6), is a right angle.

Cor. 2. If it is required to find the pole of the arc CD, draw the indefinite arc DA perpendicular to CD, and take DA equal to a quadrant; the point A will be one of the poles of the arc CD. Or, at each of the extremities C and D, draw the arcs CA and DA perpendicular to CD; the point of inter section of these arcs will be the pole required.

Cor. 3. Conversely, if the distance of the point A from each of the points C and D is equal to a quadrant, the point A will be the pole of the arc CD; and the angles ACD, ADC will be right angles.

For, let I be the center of the sphere, and draw the radii AI, CI, DI. Because the angles AIC, AID are right angles, the line AI is perpendicular to the two lines CI, DI; it is, therefore, perpendicular to their plane (Prop. IV., B. VII.). Hence the point A is the pole of the arc CD (Prop. V.); and therefore the angles ACD, ADC are right angles (Cor. 1).

Scholium. Circles may be drawn upon the surface of a sphere, with the same ease as upon a plane surface. Thus, by revolving the arc AF around the point A, the point F will describe the small circle FGH; and if we revolve the quadrant AC around the point A, the extremity C will describe the great circle CDE.

If it is required to produce the arc CD, or if it is required to draw an arc of a great circle through the two points C and D, then from the points C and D as centers, with a radius equal to a quadrant, describe two arcs intersecting each other in A. The point A will be the pole of the arc CD; and, therefore, if, from A as a center, with a radius equal to a quadrant, we describe a circle CDE, it will be a great circle passing through C and D.

If it is required to let fall a perpendicular from any point G upon the arc CD; produce CD to L, making GL equal to a quadrant; then from the pole L, with the radius GL, describe the arc GD; it will be perpendicular to CD.

PROPOSITION VI. THEOREM.

A plane, perpendicular to a diameter at its extremity, touches the sphere.

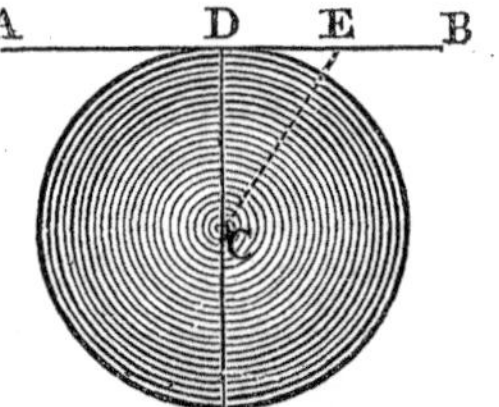

Let ADB be a plane perpendicular to the diameter DC at its extremity; then the plane ADB touches the sphere.

Let E be any point in the plane ADB, and join DE, CE. Because CD is perpendicular to the plane ADB, it is perpendicular to the line AB (Def. 1, B. VII.); hence the angle CDE is a right angle, and the line CE is greater than CD. Consequently, the point E lies without the sphere. Hence the plane ADB has only the point D in common with the sphere; it therefore touches the sphere (Def. 4). Therefore, a plane, &c.

Cor. In the same manner, it may be proved that two spheres touch each other, when the distance between their centers is equal to the sum or difference of their radii; in which case, the centers and the point of contact lie in one straight line.

PROPOSITION VII. THEOREM.

The angle formed by two arcs of great circles, is equal to the angle formed by the tangents of those arcs at the point of their intersection; and is measured by the arc of a great circle described from its vertex as a pole, and included between its sides.

Let BAD be an angle formed by two arcs of great circles; then will it be equal to the angle EAF formed by the tangents of these arcs at the point A, and it is measured by the arc DB described from the vertex A as a pole.

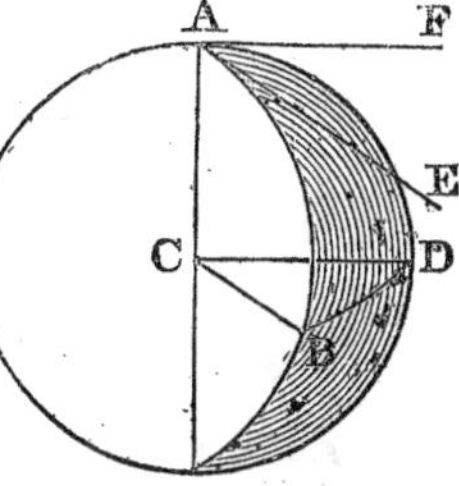

For the tangent AE, drawn in the plane of the arc AB, is perpendicular to the radius AC (Prop. IX., B. III.); also, the tangent AF, drawn in the plane of the arc AD, is perpendicular to the same radius AC. Hence the angle EAF is equal to the angle of the planes ACB, ACD (Def. 4, B. VII.), which is the same as that of the arcs AB, AD.

Also, if the arcs AB, AD are each equal to a quadrant, the lines CB, CD will be perpendicular to AC, and the angle BCD will be equal to the angle of the planes ACB, ACD; hence the arc BD measures the angle of the planes, or the angle BAD.

Cor. 1. Angles of spherical triangles may be compared with each other by means of arcs of great circles described from their vertices as poles, and included between their sides; and thus an angle can easily be made equal to a given angle.

Cor. 2. If two arcs of great circles AC, DE cut each other, the vertical angles ABE, DBC are equal; for each is equal to the angle formed by the two planes ABC, DBE. Also, the two adjacent angles ABD, DBC are together equal to two right angles.

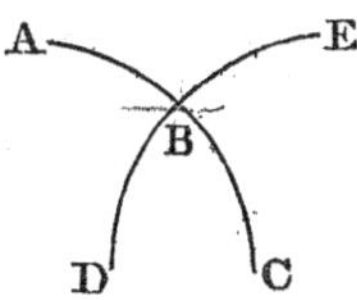

PROPOSITION VIII. THEOREM.

If from the vertices of a given spherical triangle, as poles, arcs of great circles are described, a second triangle is formed, whose vertices are poles of the sides of the given triangle.

Let ABC be a spherical triangle; and from the points A, B, C, as poles, let great circles be described intersecting each other in D, E, and F; then will the points D, E, and F be the poles of the sides of the triangle ABC.

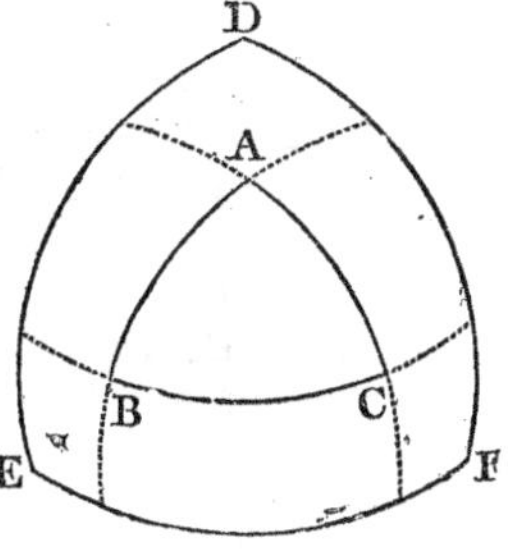

For, because the point A is the pole of the arc EF, the distance from A to E is a quadrant. Also, because the point C is the pole of the arc DE, the distance from C to E is a quadrant. Hence the point E is at a quadrant's distance from each of the points A and C; it is, therefore, the pole of the arc AC (Prop. V., Cor. 3). In the same manner, it may be proved that D is the pole of the arc BC, and F the pole of the arc AB.

Scholium. The triangle DEF is called the *polar triangle* of ABC; and so, also, ABC is the polar triangle of DEF.

Several different triangles might be formed by producing the sides DE, EF, DF; but we shall confine ourselves to the central triangle, of which the vertex D is on the same side of BC with the vertex A; E is on the same side of AC with the vertex B; and F is on the same side of AB with the vertex C.

PROPOSITION IX. THEOREM.

The sides of a spherical triangle, are the supplements of the arcs which measure the angles of its polar triangle; and conversely.

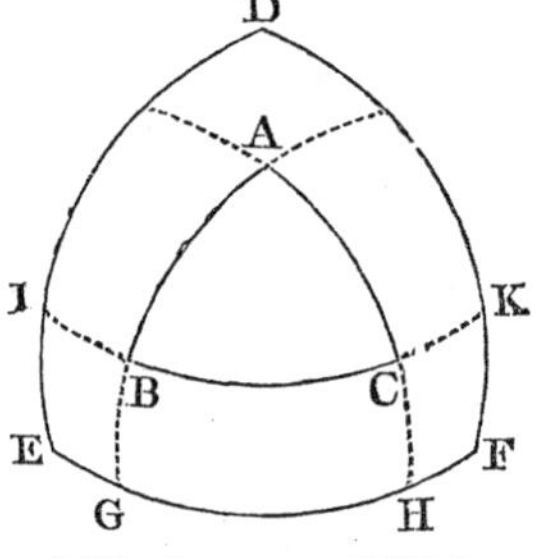

Let DEF be a spherical triangle, ABC its polar triangle; then will the side EF be the supplement of the arc which measures the angle A; and the side BC is the supplement of the arc which measures the angle D.

Produce the sides AB, AC, if necessary, until they meet EF in G and H. Then, because the point A is the pole of the arc GH, the angle A is measured by the arc GH (Prop. VII.). Also, because E is the pole of the arc AH, the arc EH is a quadrant; and, because F is the pole of AG, the arc FG is a quadrant. Hence EH and GF, or EF and GH, are together equal to a semicircumference. Therefore EF is the supplement of GH, which measures the angle A. So, also, DF is the supplement of the arc which measures the angle B; and DE is the supplement of the arc which measures the angle C.

Conversely. Because the point D is the pole of the arc BC, the angle D is measured by the arc IK. Also, because C is the pole of the arc DE, the arc IC is a quadrant; and, because B is the pole of the arc DF, the arc BK is a quadrant. Hence IC and BK, or IK and BC, are together equal to a semicircumference. Therefore BC is the supplement of IK, which measures the angle D. So, also, AC is the supplement of the arc which measures the angle E; and AB is the supplement of the arc which measures the angle F.

PROPOSITION X. THEOREM

The sum of the angles of a spherical triangle, is greater than two, and less than six right angles.

Let A, B, and C be the angles of a spherical triangle. The arcs which measure the angles A, B, and C, together with the three sides of the polar triangle, are equal to three semicircumferences (Prop. IX). But the three sides of the polar triangle are less than two semicircumferences (Prop. IV.); hence the arcs which measure the angles A, B, and C are greater than one semicircumference; and, therefore, the angles A, B, and C are greater than two right angles.

Also, because each angle of a spherical triangle is less than two right angles, the sum of the three angles must be less than six right angles.

Cor. A spherical triangle may have two, or even three, right angles; also two, or even three, obtuse angles. If a triangle have three right angles, each of its sides will be a quadrant, and the triangle is called a *quadrantal* triangle. The quadrantal triangle is contained eight times in the surface of the sphere.

PROPOSITION XI. THEOREM.

If two triangles on equal spheres are mutually equilateral, they are mutually equiangular.

Let ABC, DEF be two triangles on equal spheres, having the sides AB equal to DE, AC to DF, and BC to EF; then will the angles also be equal, each to each.

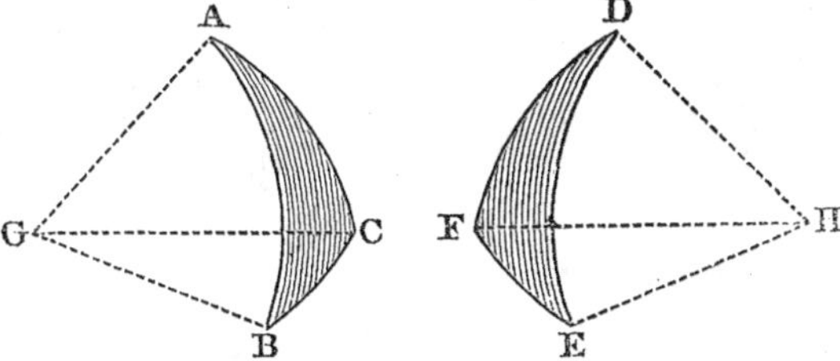

Let the centers of the spheres be G and H, and draw the radii GA, GB, GC, HD, HE, HF. A solid angle may be conceived as formed at G by the three plane angles AGB, AGC,

BGC; and another solid angle at H by the three plane angles DHE, DHF, EHF. Then, because the arcs AB, DE are equal, the angles AGB, DHE, which are measured by these arcs, are equal. For the same reason, the angles AGC, DHF are equal to each other; and, also, BGC equal to EHF

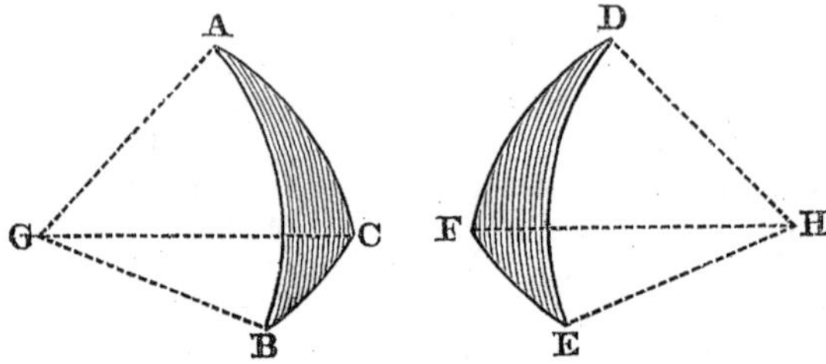

Hence G and H are two solid angles contained by three equal plane angles; therefore the planes of these equal angles are equally inclined to each other (Prop. XIX., B. VII.). That is, the angles of the triangle ABC are equal to those of the triangle DEF, viz., the angle ABC to the angle DEF, BAC to EDF, and ACB to DFE.

Scholium. It should be observed that the two triangles ABC, DEF do not admit of superposition, unless the three sides are *similarly situated* in both cases. Triangles which are mutually equilateral, but can not be applied to each other so as to coincide, are called *symmetrical* triangles.

PROPOSITION XII. THEOREM.

If two triangles on equal spheres are mutually equiangular they are mutually equilateral.

Denote by A and B two spherical triangles which are mutually equiangular, and by P and Q their polar triangles.

Since the sides of P and Q are the supplements of the arcs which measure the angles of A and B (Prop. IX.), P and Q must be mutually equilateral. Also, because P and Q are mutually equilateral, they must be mutually equiangular (Prop. XI.). But the sides of A and B are the supplements of the arcs which measure the angles of P and Q; and, therefore, A and B are mutually equilateral.

PROPOSITION XIII. THEOREM.

If two triangles on equal spheres have two sides, and the included angle of the one, equal to two sides and the included angle of the other, each to each, their third sides will be equal, and their other angles will be equal, each to each.

Let ABC, DEF be two triangles, having the side AB equal to DE, AC equal to DF, and the angle BAC equal to the angle EDF; then will the side BC be equal to EF, the angle ABC to DEF, and ACB to DFE.

If the equal sides in the two triangles are similarly situated, the triangle ABC may be applied to the triangle DEF in the same manner as in plane triangles (Prop. VI., B. I.); and the two triangles will coincide throughout. Therefore all the parts of the one triangle, will be equal to the corresponding parts of the other triangle.

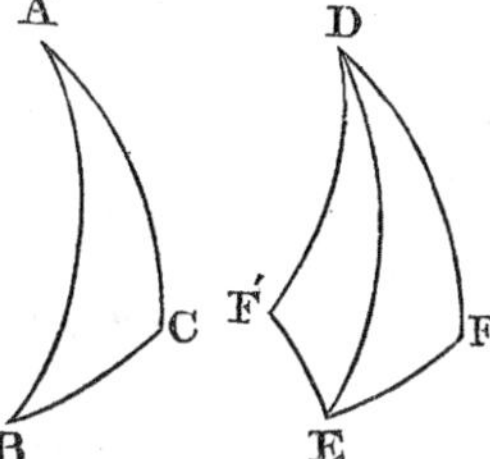

But if the equal sides in the two triangles are not similarly situated, then construct the triangle DF′E symmetrical with DFE, having DF′ equal to DF, and EF′ equal to EF. The two triangles DEF′, DEF, being mutually equilateral, are also mutually equiangular (Prop. XI.). Now the triangle ABC may be applied to the triangle DEF′, so as to coincide throughout; and hence all the parts of the one triangle, will be equal to the corresponding parts of the other triangle. Therefore the side BC, being equal to EF′, is also equal to EF; the angle ABC, being equal to DEF′, is also equal to DEF; and the angle ACB, being equal to DF′E, is also equal to DFE. Therefore, if two triangles, &c.

PROPOSITION XIV. THEOREM.

If two triangles on equal spheres have two angles, and the included side of the one, equal to two angles and the included side of the other, each to each, their third angles will be equal, and their other sides will be equal, each to each.

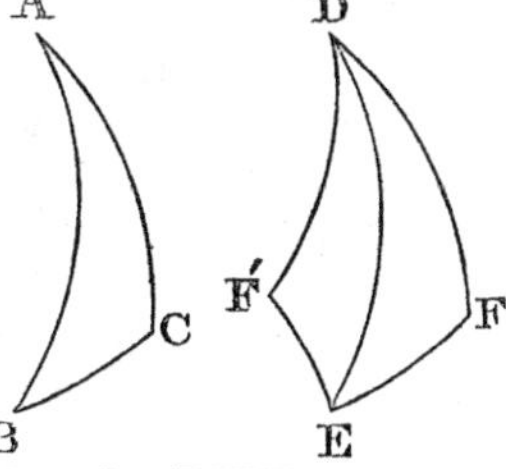

If the two triangles ABC, DEF have the angle BAC equal to the angle EDF, the angle ABC equal to DEF, and the included side AB equal to DE; the triangle ABC can be placed upon the triangle DEF, or upon its symmetrical triangle DEF′, so as to coincide. Hence the remaining parts of the triangle ABC, will be equal to the remaining parts of the triangle DEF; that is, the side AC will be equal to DF, BC to EF, and the angle ACB to the angle DFE. Therefore, if two triangles, &c.

PROPOSITION XV. THEOREM.

If two triangles on equal spheres are mutually equilateral they are equivalent.

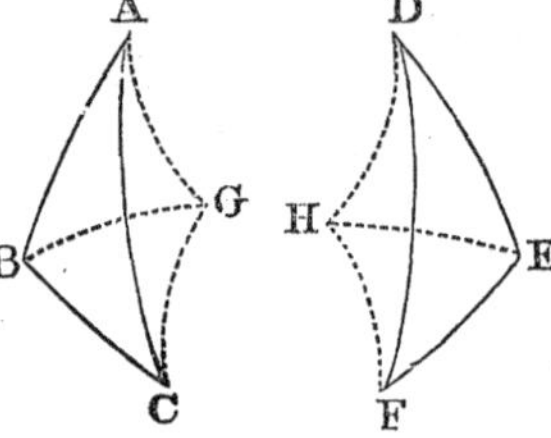

Let ABC, DEF be two triangles which have the three sides of the one, equal to the three sides of the other, each to each, viz., AB to DE, AC to DF, and BC to EF; then will the triangle ABC be equivalent to the triangle DEF.

Let G be the pole of the small circle passing through the three points A, B, C; draw the arcs GA, GB, GC; these arcs will be equal to each other (Prop. V.). At the point E, make the angle DEH equal to the angle ABG; make the arc EH equal to the arc BG; and join DH, FH.

Because, in the triangles ABG, DEH, the sides DE, EH are equal to the sides AB, BG, and the included angle DEH is equal to ABG; the arc DH is equal to AG, and the angle DHE equal to AGB (Prop. XIII.).

Now, because the triangles ABC, DEF are mutually equilateral, they are mutually equiangular (Prop. XI.); hence the angle ABC is equal to the angle DEF. Subtracting the equal angles ABG, DEH, the remainder GBC will be equal to the remainder HEF. Moreover, the sides BG, BC are equal to the sides EH, EF; hence the arc HF is equal to the arc GC, and the angle EHF to the angle BGC (Prop. XIII.).

Now the triangle DEH may be applied to the triangle ABG so as to coincide. For, place DH upon its equal BG and HE upon its equal AG, they will coincide, because the angle DHE is equal to the angle AGB; therefore the two triangles coincide throughout, and have equal surfaces. For the same reason, the surface HEF is equal to the surface GBC, and the surface DFH to the surface ACG. Hence

$$ABG+GBC-ACG=DEH+EHF-DFH;$$

or,

$$ABC=DEF;$$

that is, the two triangles ABC, DEF are equivalent. Therefore, if two triangles, &c.

Scholium. The poles G and H might be situated within the triangles ABC, DEF; in which case it would be necessary to add the three triangles ABG, GBC, ACG to form the triangle ABC; and also to add the three triangles DEH,

EHF, DFH to form the triangle DEF; otherwise the demonstration would be the same as above.

Cor. If two triangles on equal spheres, are mutually equiangular, they are equivalent. They are also equivalent, if they have two sides, and the included angle of the one, equal to two sides and the included angle of the other, each to each; or two angles and the included side of the one, equal to two angles and the included side of the other

PROPOSITION XVI. THEOREM.

In an isosceles spherical triangle, the angles opposite the equal sides are equal; and, conversely, if two angles of a spherical triangle are equal, the triangle is isosceles.

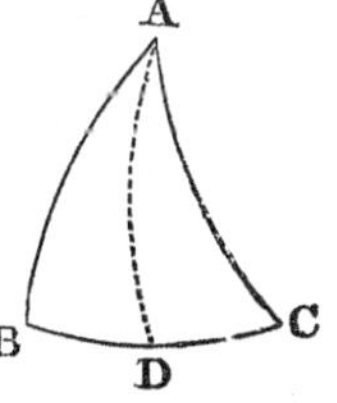

Let ABC be a spherical triangle, having the side AB equal to AC; then will the angle ABC be equal to the angle ACB.

From the point A draw the arc AD to the middle of the base BC. Then, in the two triangles ABD, ACD, the side AB is equal to AC, BD is equal to DC, and the side AD is common; hence the angle ABD is equal to the angle ACD (Prop. XI.).

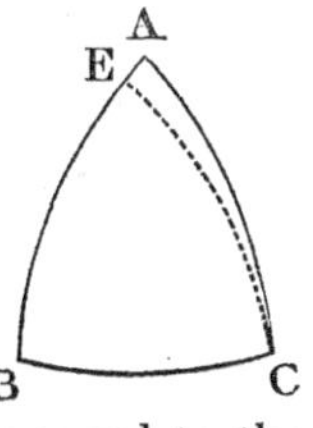

Conversely. Let the angle B be equal to the angle C; then will the side AC be equal to the side AB.

For if the two sides are not equal to each other, let AB be the greater; take BE equal to AC, and join EC. Then, in the triangles EBC, ACB, the two sides BE, BC are equal to the two sides CA, CB, and the included angles EBC, ACB are equal; hence the angle ECB is equal to the angle ABC (Prop. XIII.). But, by hypothesis, the angle ABC is equal to ACB; hence ECB is equal to ACB, which is absurd. Therefore AB is not greater than AC; and, in the same manner, it can be proved that it is not less; it is, consequently, equal to AC. Therefore, in an isosceles spherical triangle, &c.

Cor. The angle BAD is equal to the angle CAD, and the angle ADB to the angle ADC; therefore each of the last two angles is a right angle. Hence *the arc drawn from the vertex of an isosceles spherical triangle, to the middle of the base, is perpendicular to the base, and bisects the vertical angle.*

PROPOSITION XVII. THEOREM.

In a spherical triangle, the greater side is opposite the greater angle, and conversely.

Let ABC be a spherical triangle, having the angle A greater than the angle B; then will the side BC be greater than the side AC.

Draw the arc AD, making the angle BAD equal to B. Then, in the triangle ABD, we shall have AD equal to DB (Prop. XVI.); that is, BC is equal to the sum of AD and DC But AD and DC are together greater than AC (Prop. II.); hence BC is greater than AC.

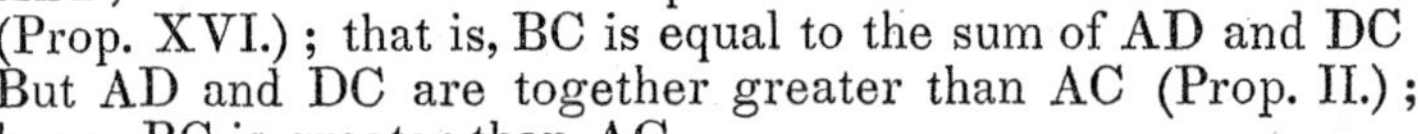

Conversely. If the side BC is greater than AC, then will the angle A be greater than the angle B. For if the angle A is not greater than B, it must be either equal to it, or less. It is not equal; for then the side BC would be equal to AC (Prop. XVI.), which is contrary to the hypothesis. Neither can it be less; for then the side BC would be less than AC, by the first case, which is also contrary to the hypothesis. Hence the angle BAC is greater than the angle ABC. Therefore, in a spherical triangle, &c.

PROPOSITION XVIII. THEOREM.

The area of a lune is to the surface of the sphere, as the angle of the lune is to four right angles.

Let ADBE be a lune, upon a sphere whose center is C, and the diameter AB; then will the area of the lune be to the surface of the sphere, as the angle DCE to four right angles, or as the arc DE to the circumference of a great circle.

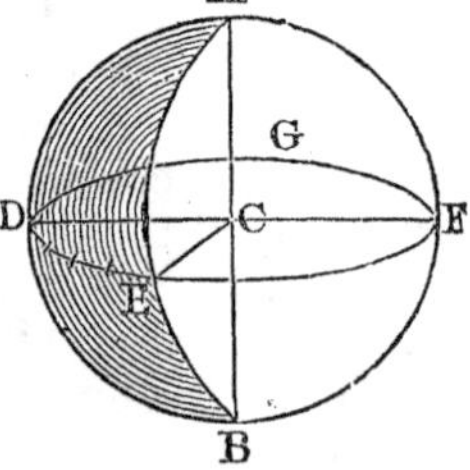

First. When the ratio of the arc to the circumference can be expressed in whole numbers.

Suppose the ratio of DE to DEFG to be as 4 to 25. Now if we divide the circumference DEFG in 25 equal parts, DE will contain 4 of those parts. If we join the pole A and the several points of division, by arcs of great circles, there will be formed on the hemisphere ADEFG, 25 triangles, all equal to each other, being mutually equilateral. The entire sphere will contain 50 of these small triangles, and the lune ADBE 8 of them. Hence the area of the lune is to the surface of the sphere, as 8 to 50, or as 4 to 25; that is, as the arc DE to the circumference.

Secondly. When the ratio of the arc to the circumference can not be expressed in whole numbers, it may be proved, as in Prop. XIV., B. III., that the lune is still to the surface of the sphere, as the angle of the lune to four right angles.

Cor. 1. On equal spheres, two lunes are to each other as the angles included between their planes.

Cor. 2. We have seen that the entire surface of the sphere is equal to eight quadrantal triangles (Prop. X., Cor.). If the area of the quadrantal triangle be represented by T, the surface of the sphere will be represented by 8T. Also, if we take the right angle for unity, and represent the angle of the lune by A, we shall have the proportion

$$\textit{area of the lune} : 8T :: A : 4.$$

Hence the area of the lune is equal to $\frac{8A \times T}{4}$, or $2A \times T$.

Cor. 3. The spherical ungula, comprehended by the planes ADB, AEB, is to the entire sphere, as the angle DCE is to four right angles. For the lunes being equal, the spherical ungulas will also be equal; hence, in equal spheres, two ungulas are to each other as the angles included between their planes.

PROPOSITION XIX. THEOREM.

If two great circles intersect each other on the surface of a hemisphere, the sum of the opposite triangles thus formed, is equivalent to a lune, whose angle is equal to the inclination of the two circles.

Let the great circles ABC, DBE intersect each other on the surface of the hemisphere BADCE; then will the sum of the opposite triangles ABD, CBE be equivalent to a lune whose angle is CBE.

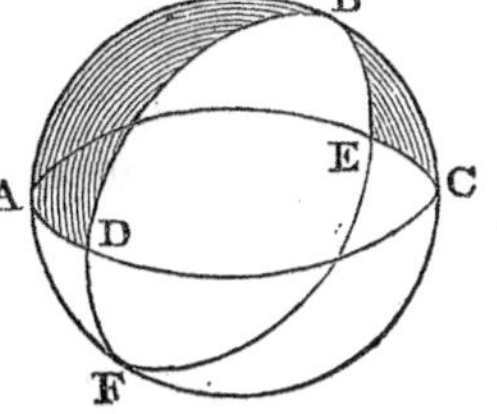

For, produce the arcs BC, BE till they meet in F; then will BCF be a semicircumference, as also ABC. Subtracting BC from each, we shall have CF equal to AB. For the same reason EF is equal to DB, and CE is equal to AD.

Hence the two triangles ABD, CFE are mutually equilateral; they are, therefore, equivalent (Prop. XV.). But the two triangles CBE, CFE compose the lune BCFE, whose angle is CBE; hence the sum of the triangles ABD, CBE is equivalent to the lune whose angle is CBE. Therefore, if two great circles, &c.

PROPOSITION XX. THEOREM.

The surface of a spherical triangle is measured by the excess of the sum of its angles above two right angles, multiplied by the quadrantal triangle.

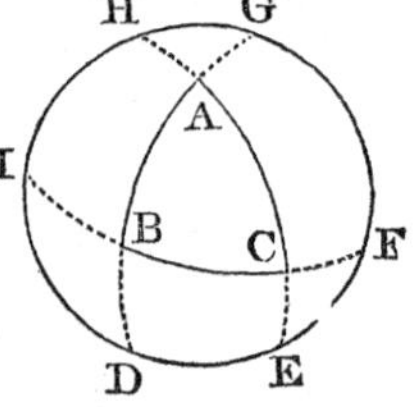

Let ABC be any spherical triangle; its surface is measured by the sum of its angles A, B, C diminished by two right angles, and multiplied by the quadrantal triangle.

Produce the sides of the triangle ABC, until they meet the great circle DEG, drawn without the triangle. The two triangles ADE, AGH are together equal to the lune whose angle is A (Prop. XIX.); and this lune is measured by $2A \times T$ (Prop. XVIII., Cor. 2). Hence we have

$$ADE+AGH=2A \times T.$$

For the same reason,

$$BFG+BDI=2B \times T;$$

also,

$$CHI+CEF=2C \times T.$$

But the sum of these six triangles exceeds the surface of the hemisphere, by twice the triangle ABC; and the hemisphere is represented by 4T; hence we have

$$4T+2ABC=2A \times T+2B \times T+2C \times T;$$

or, dividing by 2, and then subtracting 2T from each of these equals, we have

$$ABC=A \times T+B \times T+C \times T-2T,$$
$$=(A+B+C-2) \times T.$$

Hence every spherical triangle is measured by the sum of its angles diminished by two right angles, and multiplied by the quadrantal triangle.

Cor. If the sum of the three angles of a triangle is equal to three right angles, its surface will be equal to the quadrantal triangle; if the sum is equal to four right angles, the surface of the triangle will be equal to two quadrantal triangles; if the sum is equal to five right angles, the surface will be equal to three quadrantal triangles, etc.

PROPOSITION XXI. THEOREM.

The surface of a spherical polygon is measured by the sum of its angles, diminished by as many times two right angles as it has sides less two, multiplied by the quadrantal triangle.

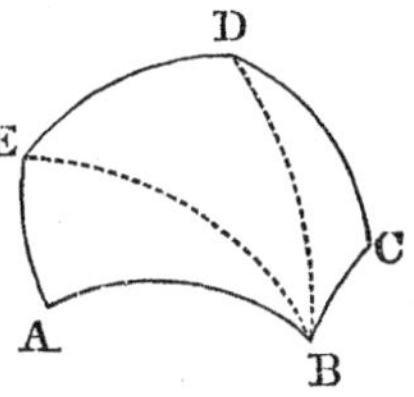

Let ABCDE be any spherical polygon. From the vertex B draw the arcs BD, BE to the opposite angles; the polygon will be divided into as many triangles as it has sides, minus two. But the surface of each triangle is measured by the sum of its angles minus two right angles, multiplied by the quadrantal triangle. Also, the sum of all the angles of the triangles, is equal to the sum of all the angles of the polygon; hence the surface of the polygon is measured by the sum of its angles, diminished by as many times two right angles as it has sides less two, multiplied by the quadrantal triangle.

Cor. If the polygon has five sides, and the sum of its an gles is equal to seven right angles, its surface will be equal to the quadrantal triangle; if the sum is equal to eight right angles, its surface will be equal to two quadrantal triangles; if the sum is equal to nine right angles, the surface will be equal to three quadrantal triangles, etc.

BOOK X.

THE THREE ROUND BODIES.

Definitions.

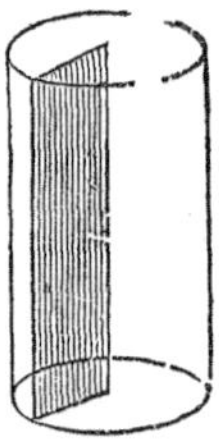

1. A *cylinder* is a solid described by the revolution of a rectangle about one of its sides, which remains fixed. The *bases* of the cylinder are the circles described by the two revolving opposite sides of the rectangle.

2. The *axis* of a cylinder is the fixed straight line about which the rectangle revolves. The opposite side of the rectangle describes the *convex surface.*

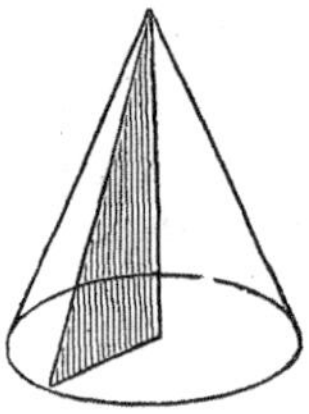

3. A *cone* is a solid described by the revolution of a right-angled triangle about one of the sides containing the right angle, which side remains fixed. The *base* of the cone is the circle described by that side containing the right angle, which revolves.

4. The *axis* of a cone is the fixed straight line about which the triangle revolves. The hypothenuse of the triangle describes the *convex surface.* The *side* of the cone is the distance from the vertex to the circumference of the base.

5. A *frustum* of a cone is the part of a cone next the base, cut off by a plane parallel to the base.

6. *Similar* cones and cylinders are those which have their axes and the diameters of their bases proportionals.

PROPOSITION I. THEOREM.

The convex surface of a cylinder is equal to the product of its altitude by the circumference of its base.

Let ACE–G be a cylinder whose base is the circle ACE and altitude AG; then will its convex surface be equal to the product of AG by the circumference ACE.

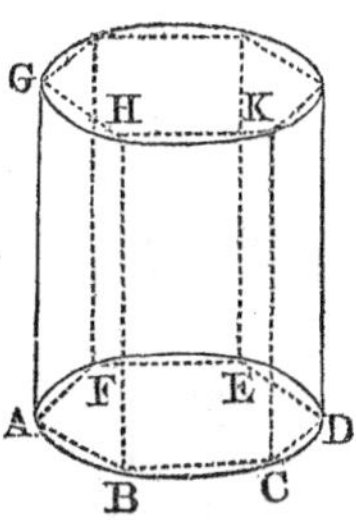

In the circle ACE inscribe the regular polygon ABCDEF; and upon this polygon let a right prism be constructed of the same altitude with the cylinder. The edges AG, BH, CK, &c., of the prism, being perpendicular to the plane of the base, will be contained in the convex surface of the cylinder. The convex surface of this prism is equal to the product of its altitude by the perimeter of its base (Prop. I., B. VIII.). Let, now, the arcs subtended by the sides AB, BC, &c., be bisected, and the number of sides of the polygon be indefinitely increased; its perimeter will approach the circumference of the circle, and will be ultimately equal to it (Prop. XI., B. VI.); and the convex surface of the prism will become equal to the convex surface of the cylinder. But whatever be the number of sides of the prism, its convex surface is equal to the product of its altitude by the perimeter of its base; hence the convex surface of the cylinder is equal to the product of its altitude by the circumference of its base.

Cor. If A represent the altitude of a cylinder, and R the radius of its base, the circumference of the base will be represented by $2\pi R$ (Prop. XIII., Cor. 2, B. VI.); and the convex surface of the cylinder by $2\pi RA$.

PROPOSITION II. THEOREM.

The solidity of a cylinder is equal to the product of its base by its altitude.

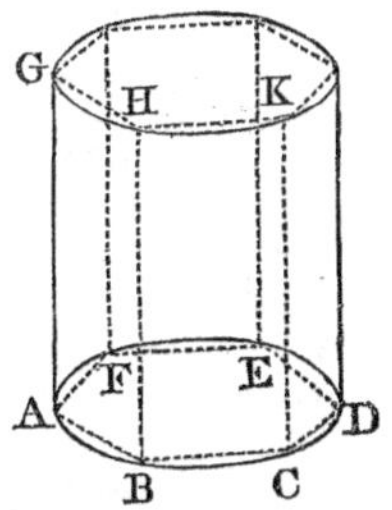

Let ACE–G be a cylinder whose base is the circle ACE and altitude AG; its solidity is equal to the product of its base by its altitude.

In the circle ACE inscribe the regular polygon ABCDEF; and upon this polygon let a right prism be constructed of the same altitude with the cylinder. The solidity of this prism is equal to the product of its base by its altitude (Prop. XI., B. VIII.). Let, now, the number of sides of the polygon be indefinitely increased; its area will become equal to that of the circle, and the solidity of the prism becomes equal to that of the cylinder. But whatever be the number of sides of the prism, its solidity is equal to the product of its base by its altitude; hence the solidity of a cylinder is equal to the product of its base by its altitude

Cor. 1. If A represent the altitude of a cylinder, and R the radius of its base, the area of the base will be represented by πR^2 (Prop. XIII., Cor. 3, B. VI.); and the solidity of the cylinder will be $\pi R^2 A$.

Cor. 2. Cylinders of the same altitude, are to each other as their bases; and cylinders of the same base, are to each other as their altitudes.

Cor. 3. Similar cylinders are to each other as the cubes of their altitudes, or as the cubes of the diameters of their bases. For the bases are as the squares of their diameters; and since the cylinders are similar, the diameters of the bases are as their altitudes (Def. 6). Therefore the bases are as the squares of the altitudes; and hence the products of the bases by the altitudes, or the cylinders themselves, will be as the cubes of the altitudes.

PROPOSITION III. THEOREM.

The convex surface of a cone is equal to the product of half its side, by the circumference of its base.

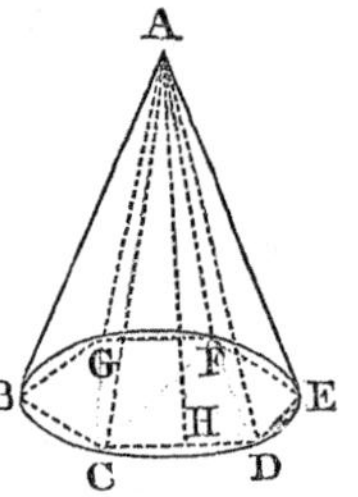

Let A–BCDEFG be a cone whose base is the circle BDEG, and its side AB; then will its convex surface be equal to the product of half its side by the circumference of the circle BDF.

In the circle BDF inscribe the regular polygon BCDEFG; and upon this polygon let a regular pyramid be constructed having A for its vertex. The edges of this pyramid will lie in the convex surface of the cone. From A draw AH perpendicular to CD, one of the sides of the polygon. The convex surface of the pyramid is equal to the product of half the slant height AH by the perimeter of its base (Prop. XIV., B. VIII.). Let, now, the arcs subtended by the sides BC, CD, &c., be bisected, and the number of sides of the polygon be indefinitely increased, its perimeter will become equal to the circumference of the circle, the slant height AH becomes equal to the side of the cone AB, and he convex surface of the pyramid becomes equal to the convex surface of the cone. But, whatever be the number of faces of the pyramid, its convex surface is equal to the product of half its slant height by the perimeter of its base; hence the convex surface of the cone, is equal to the product of half its side by the circumference of its base.

Cor. If S represent the side of a cone, and R the radius

of its base, then the circumference of the base will be represented by $2\pi R$, and the convex surface of the cone by $2\pi R \times \frac{1}{2}S$, or πRS.

PROPOSITION IV. THEOREM.

The convex surface of a frustum of a cone is equal to the product of its side, by half the sum of the circumferences of its two bases.

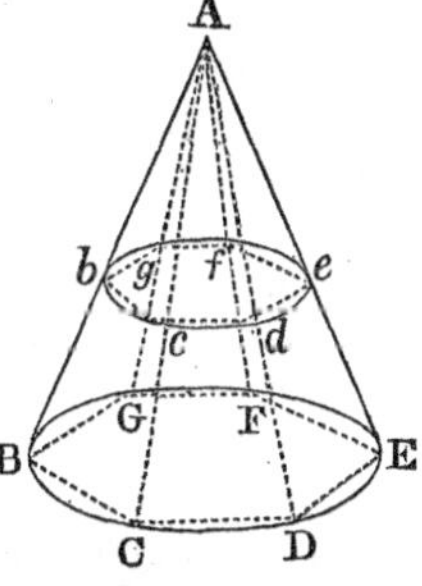

Let BDF–*bdf* be a frustum of a cone whose bases are BDF, *bdf*, and B*b* its side; its convex surface is equal to the product of B*b* by half the sum of the circumferences BDF, *bdf*.

Complete the cone A–BDF to which the frustum belongs, and in the circle BDF inscribe the regular polygon BCDEFG; and upon this polygon let a regular pyramid be constructed having A for its vertex. Then will BDF–*bdf* be a frustum of a regular pyramid, whose convex surface is equal to the product of its slant height by half the sum of the perimeters of its two bases (Prop. XIV., Cor. 1, B. VIII.). Let, now, the number of sides of the polygon be indefinitely increased, its perimeter will become equal to the circumference of the circle, and the convex surface of the pyramid will become equal to the convex surface of the cone. But, whatever be the number of faces of the pyramid, the convex surface of its frustum is equal to the product of its slant height, by half the sum of the perimeters of its two bases. Hence the convex surface of a frustum of a cone is equal to the product of its side by half the sum of the circumferences of its two bases.

Cor. It was proved (Prop. XIV., Cor. 2, B. VIII.), that the convex surface of a frustum of a pyramid is equal to the product of its slant height, by the perimeter of a section at equal distances between its two bases; hence *the convex surface of a frustum of a cone is equal to the product of its side, by the circumference of a section at equal distances between the two bases*

PROPOSITION V. THEOREM.

The solidity of a cone is equal to one third of the product of its base and altitude.

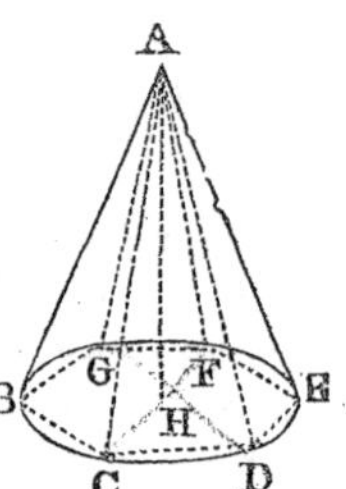

Let A–BCDF be a cone whose base is the circle BCDEFG, and AH its altitude; the solidity of the cone will be equal to one third of the product of the base BCDF by the altitude AH.

In the circle BDF inscribe a regular polygon BCDEFG, and construct a pyramid whose base is the polygon BDF, and having its vertex in A. The solidity of this pyramid is equal to one third of the product of the polygon BCDEFG by its altitude AH (Prop. XVII., B. VIII.). Let, now, the number of sides of the polygon be indefinitely increased; its area will become equal to the area of the circle, and the solidity of the pyramid will become equal to the solidity of the cone. But, whatever be the number of faces of the pyramid, its solidity is equal to one third of the product of its base and altitude; hence the solidity of the cone is equal to one third of the product of its base and altitude.

Cor. 1. Since a cone is one third of a cylinder having the same base and altitude, it follows that cones of equal altitudes are to each other as their bases; cones of equal bases are to each other as their altitudes; and similar cones are as the cubes of their altitudes, or as the cubes of the diameters of their bases.

Cor. 2. If A represent the altitude of a cone, and R the radius of its base, the solidity of the cone will be represented by $\pi R^2 \times \frac{1}{3}A$, or $\frac{1}{3}\pi R^2 A$.

PROPOSITION VI. THEOREM.

A frustum of a cone is equivalent to the sum of three cones, having the same altitude with the frustum, and whose bases are the lower base of the frustum, its upper base, and a mean proportional between them.

Let BDF–*bdf* be any frustum of a cone. Complete the cone to which the frustum belongs, and in the circle BDF inscribe the regular polygon BCDEFG; and upon this poly gon let a regular pyramid be constructed having its vertex in A. Then will BCDEFG–*bcdefg* be a frustum of a regular pyramid, whose solidity is equal to three pyramids having the same altitude with the frustum, and whose bases are the lower base of the frustum, its upper base, and a mean proportional between them (Prop. XVIII., B. VIII.). Let, now, the number of sides of the polygon be indefinitely increased, its area will become equal to the area of the circle, and the frustum of the pyramid will become the frustum of a cone Hence the frustum of a cone is equivalent to the sum of three cones, having the same altitude with the frustum, and whose bases are the lower base of the frustum, its upper base, and a mean proportional between them.

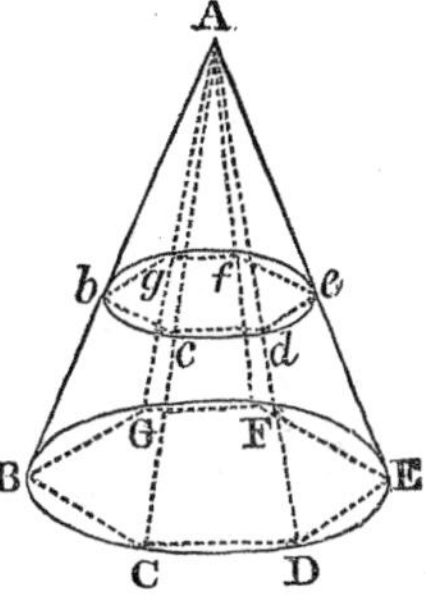

PROPOSITION VII. THEOREM.

The surface of a sphere is equal to the product of its diameter by the circumference of a great circle.

Let ABDF be the semicircle by the revolution of which the sphere is described. Inscribe in the semicircle a regular semi-polygon ABCDEF, and from the points B, C, D, E let fall the perpendiculars BG, CH, DK, EL upon the diameter AF. If, now, the polygon be revolved about AF, the lines AB, EF will describe the convex surface of two cones; and BC, CD, DE will describe the convex surface of frustums of cones.

From the center I, draw IM perpendicular to BC; also, draw MN perpendicular to AF, and BO perpendicular to CH. Let *circ.* MN represent the circumference of the circle described by the revolution of MN. Then the surface described by the revolution of BC, will be equal to BC, multiplied by *circ.* MN (Prop. IV., Cor.).

Now, the triangles IMN, BCO are similar, since their sides are perpendicular to each other (Prop. XXI., B. IV.); whence

BC : BO or GH : : IM : MN,

: : *circ.* IM : *circ.* MN.

Hence (Prop. I., B. II.),

$$BC \times circ.\ MN = GH \times circ.\ IM.$$

Therefore the surface described by BC, is equal to the altitude GH, multiplied by *circ.* IM, or the circumference of the inscribed circle.

In like manner, it may be proved that the surface described by CD is equal to the altitude HK, multiplied by the circumference of he inscribed circle; and the same may be proved of the other sides. Hence the entire surface described by ABCDEF is equal to the circumference of the inscribed circle, multiplied by the sum of the altitudes AG, GH, HK, KL, and LF; that is, the axis of the polygon.

Let, now, the arcs AB, BC, &c., be bisected, and the number of sides of the polygon be indefinitely increased, its perimeter will coincide with the circumference of the semicircle, and the perpendicular IM will become equal to the radius of the sphere; that is, the circumference of the inscribed circle will become the circumference of a great circle. Hence the surface of a sphere is equal to the product of its diameter by the circumference of a great circle.

Cor. 1. *The area of a zone is equal to the product of its altitude by the circumference of a great circle.*

For the surface described by the lines BC, CD is equal to the altitude GK, multiplied by the circumference of the inscribed circle. But when the number of sides of the polygon is indefinitely increased, the perimeter BC+CD becomes the arc BCD, and the inscribed circle becomes a great circle. Hence the area of the zone produced by the revolution of BCD, is equal to the product of its altitude GK by the circumference of a great circle.

Cor. 2. The area of a great circle is equal to the product of its circumference by half the radius (Prop. XII., B. VI.), or one fourth of the diameter; hence *the surface of a sphere is equivalent to four of its great circles.*

Cor. 3. *The surface of a sphere is equal to the convex surface of the circumscribed cylinder.*

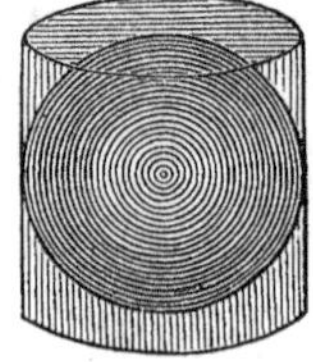

For the latter is equal to the product of its altitude by the circumference of its base. But its base is equal to a great circle of the sphere, and its altitude to the diameter; hence the convex surface of the cylinder, is equal to the product of its diameter by the circumference of a great circle, which is also the measure of the surface of a sphere.

Cor. 4. Two zones upon equal spheres, are to each other as their altitudes; and any zone is to the surface of its sphere, as the altitude of the zone is to the diameter of the sphere.

Cor. 5. Let R denote the radius of a sphere, D its diameter, C the circumference of a great circle, and S the surface of the sphere, then we shall have

$$C=2\pi R, \text{ or } \pi D \text{ (Prop. XIII., Cor. 2, B. VI.).}$$

Also, $S=2\pi R\times 2R=4\pi R^2$, or πD^2.

If A represents the altitude of a zone, its area will be

$$2\pi RA.$$

PROPOSITION VIII. THEOREM.

The solidity of a sphere is equal to one third the product of its surface by the radius.

Let ACEG be the semicircle by the revolution of which the sphere is described. Inscribe in the semicircle a regular semi-polygon ABCDEFG, and draw the radii BO, CO, DO, &c.

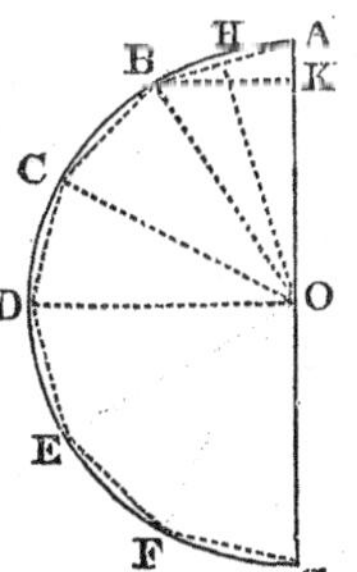

The solid described by the revolution of the polygon ABCDEFG about AG, is composed of the solids formed by the revolution of the triangles ABO, BCO, CDO, &c., about AG.

First. To find the value of the solid formed by the revolution of the triangle ABO.

From O draw OH perpendicular to AB, and from B draw BK perpendicular to AO. The two triangles ABK, BKO, in their revolution about AO, will describe two cones having a common base, viz., the circle whose radius is BK. Let *area* BK represent the area of the circle described by the revolution of BK. Then the solid described by the triangle ABO will be represented by

$$\textit{Area}\ BK\times\tfrac{1}{3}AO \text{ (Prop. V.).}$$

Now the convex surface of a cone is expressed by πRS (Prop. III., Cor.); and the base of the cone by πR^2. Hence

the convex surface : base : : $\pi RS : \pi R^2$,

: : S : R (Prop. VIII., B. II.).

But AB describes the convex surface of a cone, of which BK describes the base; hence

the surface described by AB : *area* BK : : AB : BK

: : AO : OH,

because the triangles ABK, AHO are similar. Hence

Area BK× AO= OH× surface described by AB,

or *Area* BK×$\tfrac{1}{3}$AO=$\tfrac{1}{3}$OH× surface described by AB.

But we have proved that the solid described by the triangle ABO, is equal to *area* BK$\times\frac{1}{3}$AO; it is, therefore, equal to $\frac{1}{3}$OH$\times$ surface described by AB.

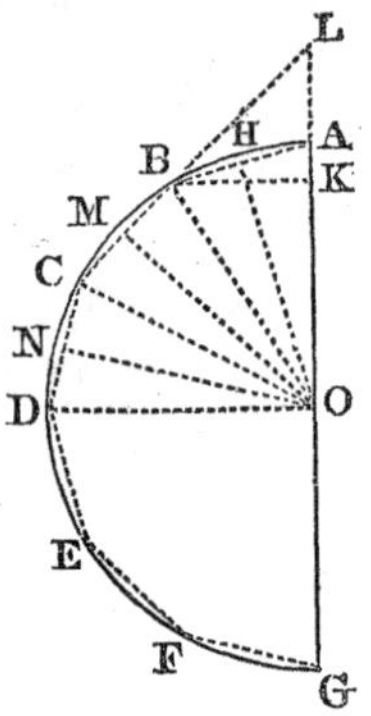

Secondly. To find the value of the solid formed by the revolution of the triangle BCO.

Produce BC until it meets AG produced in L. It is evident, from the preceding demonstration, that the solid described by the triangle LCO is equal to

$\frac{1}{3}$OM$\times$ surface described by LC;

and the solid described by the triangle LBO is equal to

$\frac{1}{3}$OM$\times$ surface described by LB;

hence the solid described by the triangle BCO is equal to

$\frac{1}{3}$OM$\times$ surface described by BC.

In the same manner, it may be proved that the solid described by the triangle CDO is equal to

$\frac{1}{3}$ON$\times$ surface described by CD;

and so on for the other triangles. But the perpendiculars OH, OM, ON, &c., are all equal; hence the solid described by the polygon ABCDEFG, is equal to the surface described by the perimeter of the polygon, multiplied by $\frac{1}{3}$OH.

Let, now, the number of sides of the polygon be indefinitely increased, the perpendicular OH will become the radius OA, the perimeter ACEG will become the semi-circumference ADG, and the solid described by the polygon becomes a sphere; hence the solidity of a sphere is equal to one third of the product of its surface by the radius.

Cor. 1. *The solidity of a spherical sector is equal to the product of the zone which forms its base, by one third of its radius.*

For the solid described by the revolution of BCDO is equal to the surface described by BC+CD, multiplied by $\frac{1}{3}$OM. But when the number of sides of the polygon is indefinitely increased, the perpendicular OM becomes the radius OB, the quadrilateral BCDO becomes the sector BDO, and the solid described by the revolution of BCDO becomes a spherical sector. Hence the solidity of a spherical sector is equal to the product of the zone which forms its base, by one third of its radius.

Cor. 2. Let R represent the radius of a sphere, D its diameter, S its surface, and V its solidity, then we shall have

$$S=4\pi R^2 \text{ or } \pi D^2 \text{ (Prop. VII., Cor. 5).}$$

Also, $V=\frac{1}{3}R\times S=\frac{4}{3}\pi R^3$ or $\frac{1}{6}\pi D^3$;

hence *the solidities of spheres are to each other as the cubes of their radii*

If we put A to represent the altitude of the zone which forms the base of a sector, then the solidity of the sector will be represented by

$$2\pi RA \times \tfrac{1}{3}R = \tfrac{2}{3}\pi R^2 A.$$

Cor. 3. *Every sphere is two thirds of the circumscribed cylinder.*

For, since the base of the circumscribed cylinder is equal to a great circle, and its altitude to a diameter, the solidity of the cylinder is equal to a great circle, multiplied by the diameter (Prop. II.). But the solidity of a sphere is equal to four great circles, multiplied by one third of the radius; or one great circle, multiplied by $\tfrac{4}{3}$ of the radius, or $\tfrac{2}{3}$ of the diameter. Hence a sphere is two thirds of the circumscribed cylinder.

PROPOSITION IX. THEOREM.

A spherical segment with one base, is equivalent to half of a cylinder having the same base and altitude, plus a sphere whose diameter is the altitude of the segment.

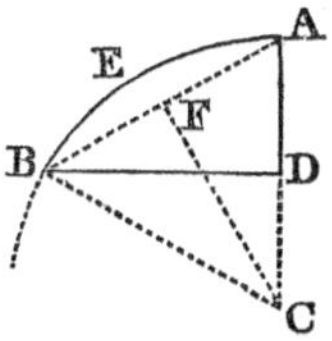

Let BD be the radius of the base of the segment, AD its altitude, and let the segment be generated by the revolution of the circular half segment AEBD about the axis AC. Join CB, and from the center C draw CF perpendicular to AB.

The solid generated by the revolution of the segment AEB, is equal to the difference of the solids generated by the sector ACBE, and the triangle ACB. Now, the solid generated by the sector ACBE is equal to

$$\tfrac{2}{3}\pi CB^2 \times AD \text{ (Prop. VIII., Cor. 2).}$$

And the solid generated by the triangle ACB, by Prop. VIII., is equal to $\tfrac{1}{3}CF$, multiplied by the convex surface described by AB, which is $2\pi CF \times AD$ (Prop. VII.), making for the solid generated by the triangle ACB,

$$\tfrac{2}{3}\pi CF^2 \times AD.$$

Therefore the solid generated by the segment AEB, is equal

to $\qquad \tfrac{2}{3}\pi AD \times (CB^2 - CF^2)$,

or $\qquad \tfrac{2}{3}\pi AD \times BF^2$;

that is, $\qquad \tfrac{1}{6}\pi AD \times AB^2$,

because $CB^2 - CF^2$ is equal to BF^2, and BF^2 is equal to one fourth of AB^2.

Now the cone generated by the triangle ABD is equal to

$$\tfrac{1}{3}\pi AD \times BD^2 \text{ (Prop. V., Cor. 2).}$$

Therefore the spherical segment in question, which is the sum of the solids described by AEB and ABD, is equal to

$$\tfrac{1}{6}\pi AD(2BD^2+AB^2);$$

that is,

$$\tfrac{1}{6}\pi AD(3BD^2+AD^2),$$

because AB^2 is equal to BD^2+AD^2.

This expression may be separated into the two parts

$$\tfrac{1}{2}\pi AD\times BD^2,$$

and

$$\tfrac{1}{6}\pi AD^3.$$

The first part represents the solidity of a cylinder having the same base with the segment and half its altitude (Prop. II.); the other part represents a sphere, of which AD is the diameter (Prop. VIII., Cor. 2). Therefore, a spherical segment, &c.

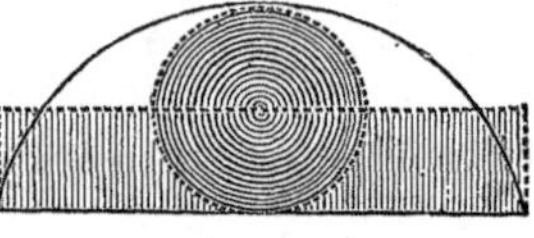

Cor. The solidity of the spherical segment of two bases, generated by the revolution of BCDE about the axis AD, may be found by subtracting that of the segment of one base generated by ABE, from that of the segment of one base generated by ACD.

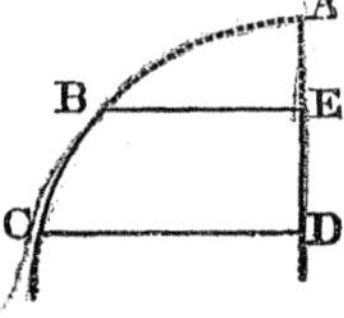

CONIC SECTIONS.

There are three curves whose properties are extensively applied in Astronomy, and many other branches of science, which, being the sections of a cone made by a plane in different positions, are called the *conic sections*. These are

The Parabola,
The Ellipse, and
The Hyperbola.

PARABOLA.

Definitions.

1. A *parabola* is a plane curve, every point of which is equally distant from a fixed point, and a given straight line.

2. The fixed point is called the *focus* of the parabola and the given straight line is called the *directrix*.

Thus, if F be a fixed point, and BC a given line, and the point A move about F in such a manner, that its distance from F is always equal to the perpendicular distance from BC, the point A will describe a parabola, of which F is the focus, and BC the directrix.

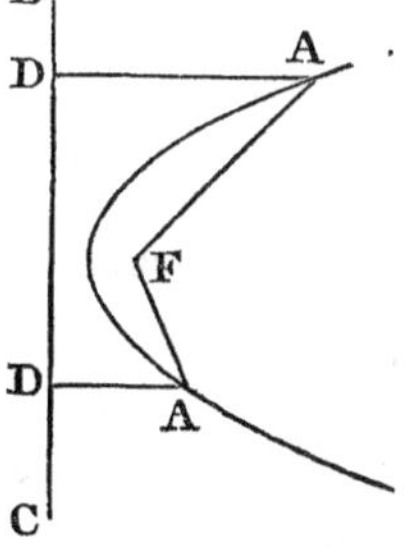

3. A *diameter* is a straight line drawn through any point of the curve perpendicular to the directrix. The *vertex* of the diameter is the point in which it cuts the curve.

Thus, through any point of the curve, as A, draw a line DE perpendicular to the directrix BC; DE is a diameter of the parabola, and the point A is the vertex of this diameter.

4. The *axis* of the parabola is the diameter which passes through the focus; and the point in which it cuts the curve is called the *principal vertex*.

Thus, draw a diameter of the parabola, GH, through the

focus F; GH is the axis of the parabola, and the point V, where the axis cuts the curve, is called the principal vertex of the parabola, or simply the vertex.

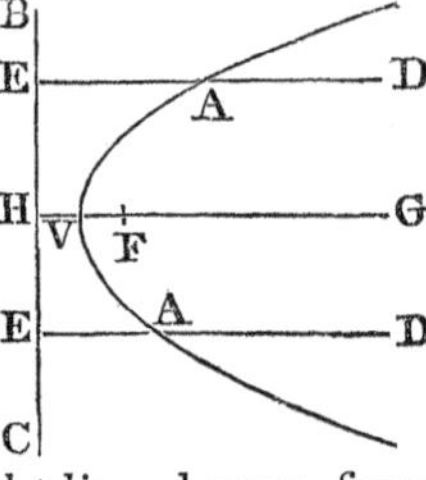

It is evident from Def. 1, that the line FH is bisected in the point V.

5. A *tangent* is a straight line which meets the curve, but, being produced, does not cut it.

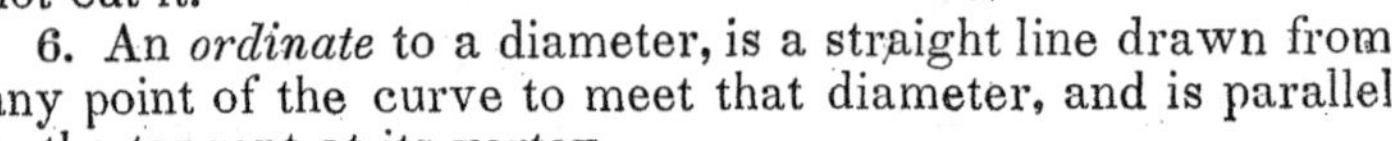

6. An *ordinate* to a diameter, is a straight line drawn from any point of the curve to meet that diameter, and is parallel to the tangent at its vertex.

Thus, let AC be a tangent to the parabola at B, the vertex of the diameter BD. From any point E of the curve, draw EGH parallel to AC; then is EG an ordinate to the diameter BD.

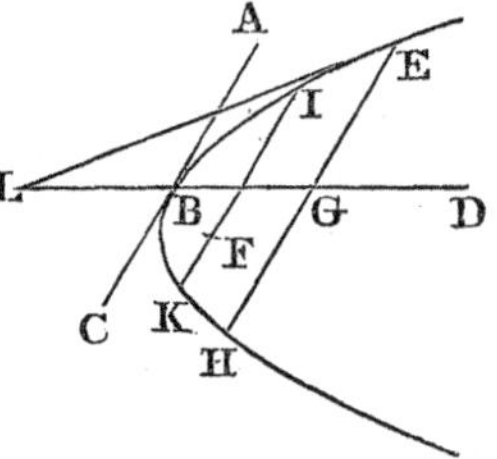

It is proved in Prop. IX., that EG is equal to GH; hence the entire line EH is called a *double ordinate.*

7. An *abscissa* is the part of a diameter intercepted between its vertex and an ordinate.

Thus, BG is the abscissa of the diameter BD, corresponding to the ordinate EG.

8. A *subtangent* is that part of a diameter intercepted between a tangent and ordinate to the point of contact.

Thus, let EL, a tangent to the curve at E, meet the diameter BD in the point L; then LG is the subtangent of BD, corresponding to the point E.

9. The *parameter* of a diameter is the double ordinate which passes through the focus.

Thus, through the focus F, draw IK parallel to the tangent AC; then is IK the parameter of the diameter BD.

10. The parameter of the axis is called the principal parameter, or *latus rectum.*

11. A *normal* is a line drawn perpendicular to a tangent from the point of contact, and terminated by the axis.

12. A *subnormal* is the part of the axis intercepted between the normal, and the corresponding ordinate.

Thus, let AB be a tangent to the parabola at any point A. From A draw AC perpendicular to AB; draw, also, the ordinate AD. Then AC is the normal, and DC is the subnormal corresponding to the point A.

PROPOSITION I. PROBLEM.

To describe a parabola.

Let BC be a ruler laid upon a plane, and let DEG be a square. Take a thread equal in length to EG, and attach one extremity at G, and the other at some point as F. Then slide the side of the square DE along the ruler BC, and, at the same time, keep the thread continually tight by means of the pencil A; the pencil will describe one part of a parabola, of which F is the focus, and BC the directrix. For, in every position of the square,

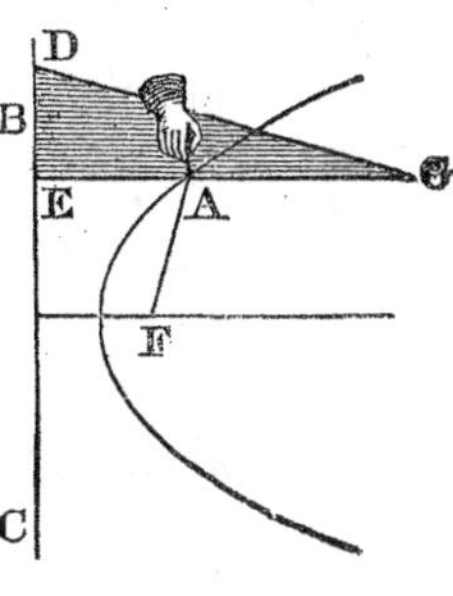

$$AF+AG=AE+AG,$$

and hence

$$AF=AE;$$

that is, the point A is always equally distant from the focus F and directrix BC.

If the square be turned over, and moved on the other side of the point F, the other part of the same parabola may be described.

PROPOSITION II. THEOREM.

A tangent to the parabola bisects the angle formed at the point of contact, by a perpendicular to the directrix, and a line drawn to the focus.

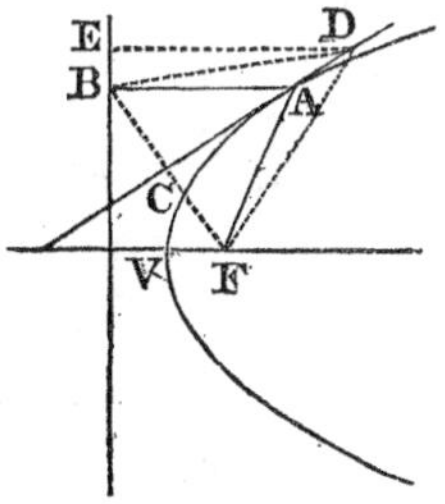

Let A be any point of the parabola AV, from which draw the line AF to the focus, and AB perpendicular to the directrix, and draw AC bisecting the angle BAF; then will AC be a tangent to the curve at the point A.

For, if possible, let the line AC meet the curve in some other point as D. Join DF, DB, and BF; also, draw DE perpendicular to the directrix.

Since, in the two triangles ACB, ACF, AF is equal to AB (Def. 1), AC is common to both triangles, and the angle CAB is, by supposition, equal to the angle CAF; therefore CB is equal to CF, and the angle ACB to the angle ACF.

Again, in the two triangles DCB, DCF, because BC is equal to CF, the side DC is common to both triangles, and the angle DCB is equal to the angle DCF; therefore DB is equal to DF. But DF is equal to DE (Def. 1); hence DB is equal to DE, which is impossible (Prop. XVII., B. I.). Therefore the line AC does not meet the curve in D; and in the same manner it may be proved that it does not meet the curve in any other point than A; consequently it is a tangent to the parabola. Therefore, a tangent, &c.

Cor. 1. Since the angle FAB continually increases as the point A moves toward V, and at V becomes equal to two right angles, *the tangent at the principal vertex is perpendicular to the axis.* The tangent at the vertex V is called the *vertical tangent.*

Cor. 2. Since an ordinate to any diameter is parallel to the tangent at its vertex, an ordinate to the axis is perpendicular to the axis.

PROPOSITION III. THEOREM.

The latus rectum is equal to four times the distance from the focus to the vertex.

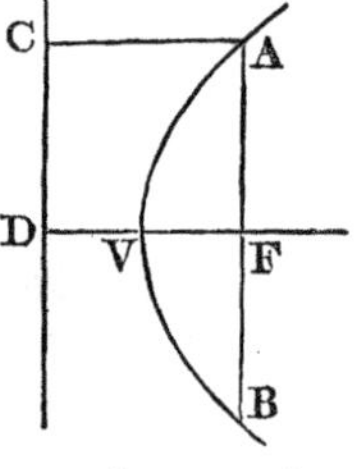

Let AVB be a parabola, of which F is the focus, and V the principal vertex; then the latus rectum AFB will be equal to four times FV.

Let CD be the directrix, and let AC be drawn perpendicular to it; then, according to Def. 1, AF is equal to AC or DF, because ACDF is a parallelogram. But DV is equal to VF; that is, DF is equal to twice VF. Hence AF is equal to twice VF. In the same manner it may be proved that BF is equal to twice VF; consequently AB is equal to four times VF. Therefore, the latus rectum, &c.

PROPOSITION IV. THEOREM.

If a tangent to the parabola cut the axis produced, the points of contact and of intersection are equally distant from the focus.

Let AB be a tangent to the parabola GAH at the point A, and let it cut the axis produced in B; also, let AF be drawn to the focus; then will the line AF be equal tc BF.

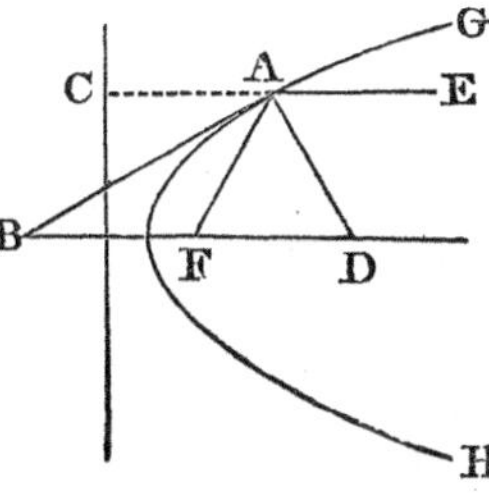

Draw AC perpendicular to the directrix; then, since AC is parallel to BF, the angle BAC is equal to ABF. But the angle BAC is equal to BAF (Prop. II.); hence the angle ABF is equal to BAF, and, consequently, AF is equal to BF. Therefore, if a tangent, &c.

Cor. 1. Let the normal AD be drawn. Then, because BAD is a right angle, it is equal to the sum of the two angles ABD ADB, or to the sum of the two angles BAF, ADB. Take away the common angle BAF, and we have the angle DAF equal to ADF. Hence the line AF is equal to FD. Therefore, *if a circle be described with the center* F, *and radius* FA, *it will pass through the three points* B, A, D.

Cor. 2. *The normal bisects the angle made by the diameter at the point of contact, with the line drawn from that point to the focus.*

For, because BD is parallel to CE, the alternate angles ADF, DAE are equal. But the angle ADF has been proved equal to DAF; hence the angles DAF, DAE are equal to each other.

Scholium. It is a law in Optics, that the angle made by a ray of reflected light with a perpendicular to the reflecting surface, is equal to the angle which the incident ray makes with the same perpendicular. Hence, if GAH represent a concave parabolic mirror, a ray of light falling upon it in the direction EA would be reflected to F. The same would be true of all rays parallel to the axis. Hence the point F, in which all the rays would intersect each other, is called the *focus*, or *burning point.*

PROPOSITION V. THEOREM.

The subtangent to the axis is bisected by the vertex.

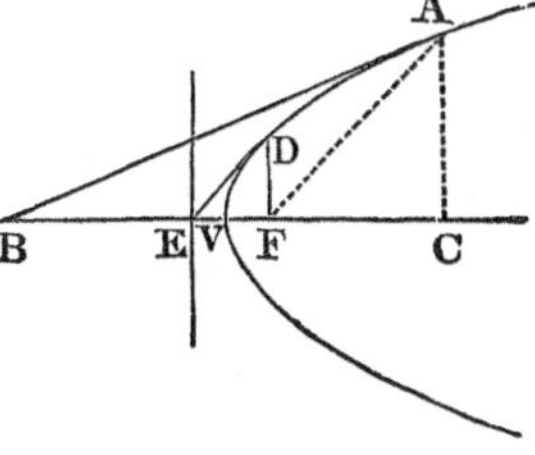

Let AB be a tangent to the parabola ADV at the point A, and AC an ordinate to the axis; then will BC be the subtangent, and it will be bisected at the vertex V.

For BF is equal to AF (Prop. IV.); and AF is equal to CE, which is the distance of the point A from the directrix. But CE is equal to the sum of CV and VE, or CV and VF. Hence BF, or

BV+VF is equal to CV+VF; that is, BV is equal to CV Therefore, the subtangent, &c.

Cor. 1. Hence the tangent at D, the extremity of the latus rectum, meets the axis in E, the same point with the directrix. For, by Def. 8, EF is the subtangent corresponding to the tangent DE.

Cor. 2. Hence, if it is required to draw a tangent to the curve at a given point A, draw the ordinate AC to the axis. Make BV equal to VC; join the points B, A, and the line BA will be the tangent required.

PROPOSITION VI. THEOREM.

The subnormal is equal to half the latus rectum.

Let AB be a tangent to the parabola AV at the point A, let AC be the ordinate, and AD the normal from the point of contact; then CD is the subnormal, and is equal to half the latus rectum.

For the distance of the point A from the focus, is equal to its distance from the directrix, which is equal to VF+VC, or 2VF+FC; that is,

$$FA=2VF+FC,$$

or

$$2VF=FA-FC.$$

Also, CD is equal to FD−FC, which is equal to FA−FC (Prop. IV., Cor. 1). Hence CD is equal to 2VF, which is equal to half the latus rectum (Prop. III.). Therefore, the subnormal, &c.

PROPOSITION VII. THEOREM.

If a perpendicular be drawn from the focus to any tangent, the point of intersection will be in the vertical tangent.

Let AB be any tangent to the parabola AV, and FC a perpendicular let fall from the focus upon AB; join VC; then will the line VC be a tangent to the curve at the vertex V.

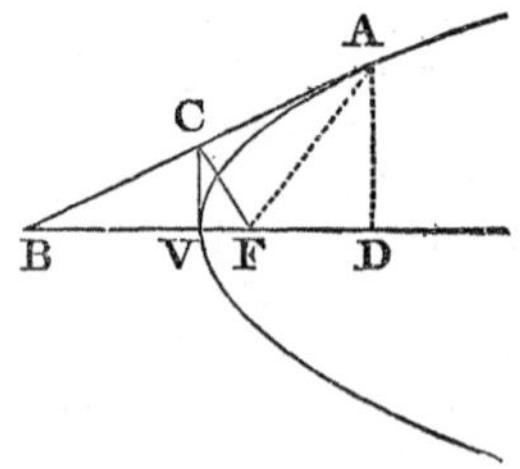

Draw the ordinate AD to the axis Since FA is equal to FB (Prop. IV.), and FC is drawn perpendicular to AB, it divides the triangle AFB into two equal parts, and, therefore, AC is equal to BC. But BV is equal to VD (Prop. V.); hence

$$BC : CA :: BV : VD,$$

and, therefore, CV is parallel to AD (Prop. XVI., B. IV.). But AD is perpendicular to the axis BD; hence CV is also perpendicular to the axis, and is a tangent to the curve at the point V (Prop. II., Cor. 1). Therefore, if a perpendicular, &c.

Cor. 1. Because the triangles FVC, FCA are similar, we have

$$FV : FC :: FC : FA;$$

that is, *the perpendicular from the focus upon any tangent, is a mean proportional between the distances of the focus from the vertex, and from the point of contact.*

Cor. 2. It is obvious that $FV : FA :: FC^2 : FA^2$.

Cor. 3. From Cor. 1, we have

$$FC^2 = FV \times FA.$$

But FV remains constant for the same parabola; therefore *the distance from the focus to the point of contact, varies as the square of the perpendicular upon the tangent.*

PROPOSITION VIII. THEOREM.

The square of an ordinate to the axis, is equal to the product of the latus rectum by the corresponding abscissa.

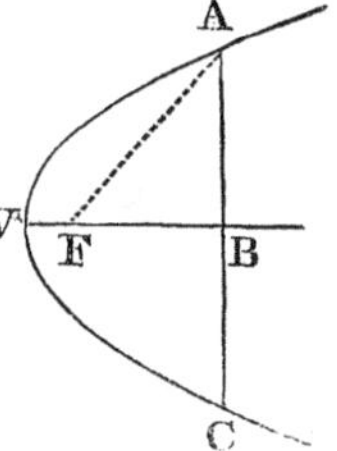

Let AVC be a parabola, and A any point of the curve. From A draw the ordinate AB; then is the square of AB equal to the product of VB by the latus rectum.

For AB^2 is equal to $AF^2 - FB^2$.

But AF is equal to VB+VF, and FB is equal to VB−VF.

Hence $AB^2 = (VB+VF)^2 - (VB-VF)^2$, which, according to Prop. IX., Cor., B. IV., is equal to

$$4VB \times VF,$$

or $VB \times$ the latus rectum (Prop. III.).

Therefore, the square, &c.

Cor. 1. Since the latus rectum is constant for the same parabola, *the squares of ordinates to the axis, are to each other as their corresponding abscissas.*

Cor. 2. The preceding demonstration is equally applicable to ordinates on either side of the axis; hence AB is equal to BC, and AC is called a *double ordinate.* The curve is symmetrical with respect to the axis, and the whole parabola is bisected by the axis.

PROPOSITION IX. THEOREM.

The square of an ordinate to any diameter, is equal to four times the product of the corresponding abscissa, by the distance from the vertex of that diameter to the focus.

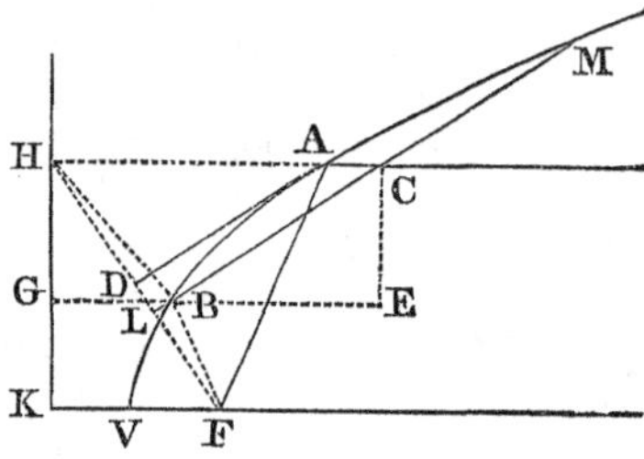

Let AD be a tangent to the parabola VAM at the point A; through A draw the diameter HAC, and through any point of the curve, as B, draw BC parallel to AD; draw also AF to the focus; then will the square of BC be equal to 4AF×AC.

Draw CE parallel, and EBG perpendicular to the directrix HK; and join BH, BF, HF. Also, produce CB to meet HF in L.

Because the right-angled triangles FHK, HCL are similar, and AD is parallel to CL, we have

$$HF : FK :: HC : HL$$
$$:: AC : DL.$$

Hence (Prop. I., B. II.),

$$HF \times DL = FK \times AC,$$

or $2HF \times DL = 2FK \times AC$, or $4VF \times AC$.

But $2HF \times DL = HL^2 - LF^2$ (Prop. X., B. IV.)

$$= HB^2 - BF^2$$
$$= HG^2 \text{ or } CE^2.$$

Hence CE^2 is equal to $4VF \times AC$.

Also, because the triangles BCE, AFD are similar, we have

$$CE : CB :: DF : AF.$$

Therefore $CE^2 : CB^2 :: DF^2 : AF^2$ (Prop. X., B. II.)

$$:: VF : AF \text{ (Prop. VII., Cor. 2)}$$
$$:: 4VF \times AC : 4AF \times AC.$$

But the two antecedents of this proportion have been proved to be equal; hence the consequents are equal, or

$$BC^2 = 4AF \times AC.$$

Therefore, the square of an ordinate, &c.

Cor. In like manner it may be proved that the square of CM is equal to 4AF×AC. Hence BC is equal to CM; and since the same may be proved for any ordinate, it follows that *every diameter bisects its double ordinates.*

PROPOSITION X. THEOREM.

The parameter of any diameter, is equal to four times the distance from its vertex to the focus.

Let BAD be a parabola, of which F is the focus, AC is any diameter, and BD its parameter; then is BD equal to four times AF.

Draw the tangent AE; then, since AEFC is a parallelogram, AC is equal to EF, which is equal to AF (Prop. IV.).

Now, by Prop. IX., BC^2 is equal to $4AF \times AC$; that is, to $4AF^2$. Hence BC is equal to twice AF, and BD is equal to four times AF. Therefore, the parameter of any diameter, &c.

Cor. Hence *the square of an ordinate to a diameter, is equal to the product of its parameter by the corresponding abscissa.*

PROPOSITION XI. THEOREM.

If a cone be cut by a plane parallel to its side, the section is a parabola.

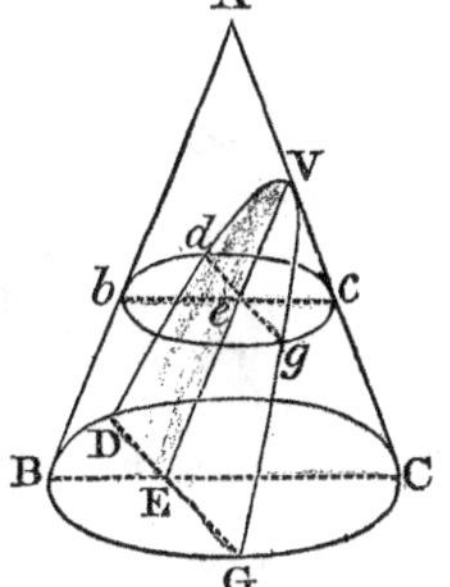

Let ABGCD be a cone cut by a plane VDG parallel to the slant side AB; then will the section DVG be a parabola.

Let ABC be a plane section through the axis of the cone, and perpendicular to the plane VDG; then VE, which is their common section, will be parallel to AB. Let *bgcd* be a plane parallel to the base of the cone; the intersection of this plane with the cone will be a circle. Since the plane ABC divides the cone into two equal parts, BC is a diameter of the circle BGCD, and *bc* is a diameter of the circle *bgcd*. Let DEG, *deg* be the common sections of the plane VDG with the planes BGCD, *bgcd* respectively. Then DG is perpendicular to the plane ABC, and, consequently, to the lines VE, BC. For the same reason, *dg* is perpendicular to the two lines VE, *bc*.

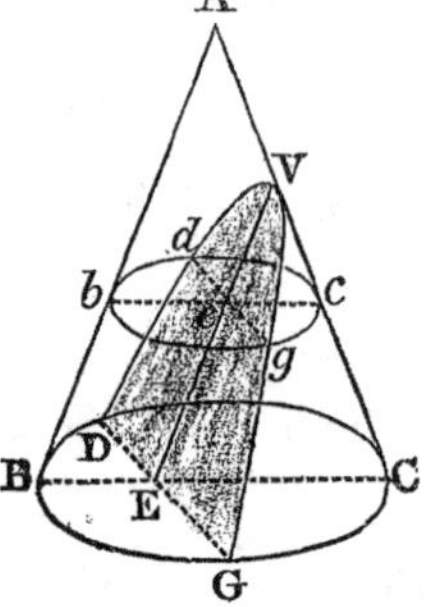

Now, since be is parallel to BE, and bB to eE, the figure bBEe is a parallelogram, and be is equal to BE. But because the triangles Vec, VEC are similar, we have

$$ec : \mathrm{EC} :: \mathrm{V}e : \mathrm{VE};$$

and multiplying the first and second terms of this proportion by the equals be and BE, we have

$$be \times ec : \mathrm{BE} \times \mathrm{EC} :: \mathrm{V}e : \mathrm{VE}.$$

But since bc is a diameter of the circle $bgcd$, and de is perpendicular to bc (Prop. XXII., Cor., B. IV.),

$$be \times ec = de^2.$$

For the same reason, $\mathrm{BE} \times \mathrm{EC} = \mathrm{DE}^2$. Substituting these values of $be \times ec$ and $\mathrm{BE} \times \mathrm{EC}$ in the preceding proportion, we have

$$de^2 : \mathrm{DE}^2 :: \mathrm{V}e : \mathrm{VE};$$

that is, the squares of the ordinates are to each other as the corresponding abscissas; and hence the curve is a parabola, whose axis is VE (Prop. VIII., Cor. 1.). Hence the parabola is called a *conic section*, as mentioned on page 177.

PROPOSITION XII. THEOREM.

Every segment of a parabola is two thirds of its circumscribing rectangle.

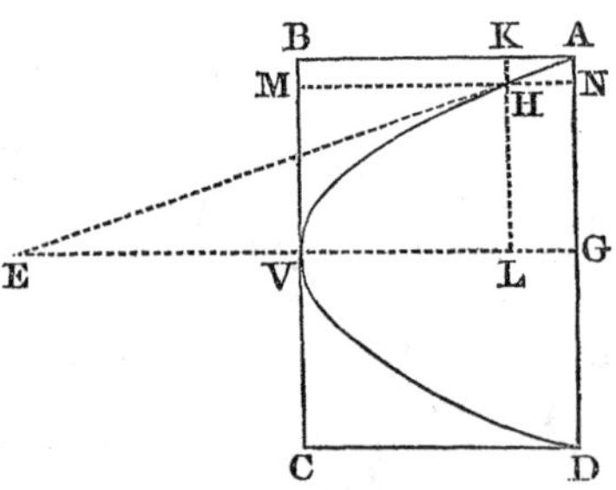

Let AVD be a segment of a parabola cut off by the straight line AD perpendicular to the axis; the area of AVD is two thirds of the circumscribing rectangle ABCD.

Draw the line AE touching the parabola at A, and meeting the axis produced in E; and take a point H in the curve, so near to A that the tangent and curve may be regarded as coinciding. Through H draw KL perpendicular, and MN parallel to the axis. Then the

$$\text{rectangle AL} : \text{rectangle AM} :: \mathrm{AG} \times \mathrm{GL} : \mathrm{AB} \times \mathrm{AN}$$
$$:: \mathrm{AG} \times \mathrm{GE} : \mathrm{AB} \times \mathrm{AG}$$
$$:: \mathrm{GE} \quad \mathrm{AB},$$

because GL or NH : AN : : GE : AG. But GE is equal to twice GV or AB (Prop. V.); hence

AL : AM : : 2 : 1;

that is, AL is double of AM.

Hence the portion of the parabola included between two ordinates indefinitely near, is double the corresponding portion of the external space ABV. Therefore, since the same is true for every point of the curve, the whole space AVG is double the space ABV. Whence AVG is two thirds of ABVG; and the segment AVD is two thirds of the rectangle ABCD. Therefore, every segment, &c.

ELLIPSE.

Definitions.

1. An *ellipse* is a plane curve, in which the sum of the distances of each point from two fixed points, is equal to a given line.

2. The two fixed points are called the *foci.*

Thus, if F, F′ are two fixed points, and if the point D moves about F in such a manner that the *sum* of its distances from F and F′ is always the same, the point D will describe an ellipse, of which F and F′ are the foci.

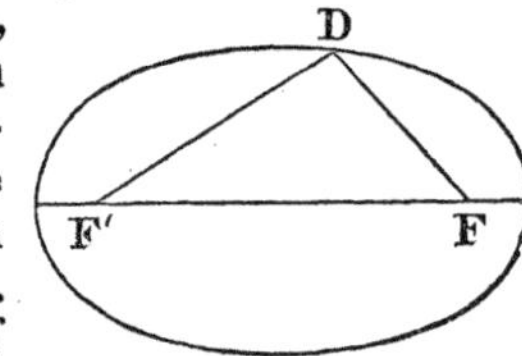

3. The *center* is the middle point of the straight line joining the foci.

4. The *eccentricity* is the distance from the center to either focus.

Thus, let ABA′B′ be an ellipse, F and F′ the foci. Draw the line FF′ and bisect it in C. The point C is the center of the ellipse; and CF or CF′ is the eccentricity.

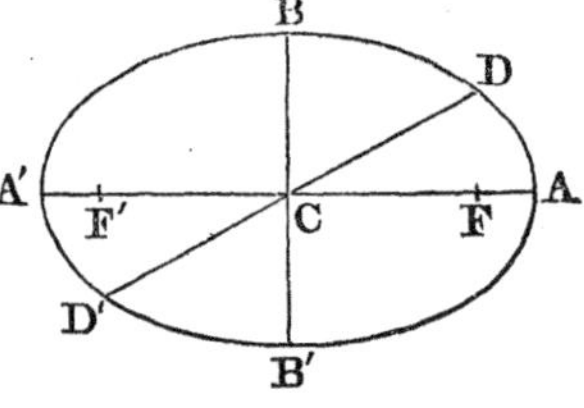

5. A *diameter* is a straight line drawn through the center, and terminated both ways by the curve.

6. The extremities of a diameter are called its *vertices.*

Thus, through C draw any straight line DD′ terminated by the curve; DD′ is a diameter of the ellipse; D and D′ are its vertices.

7. The *major axis* is the diameter which passes through the foci; and its extremities are called the *principal vertices.*

8. The *minor axis* is the diameter which is perpendicular to the major axis.

Thus, produce the line FF′ to meet the curve in A and A′; and through C draw BB′ perpendicular to AA′; then is AA′ the major axis, and BB′ the minor axis.

9. A *tangent* is a straight line which meets the curve, but, being produced, does not cut it.

10. An *ordinate* to a diameter, is a straight line drawn from any point of the curve to the diameter, parallel to the tangent at one of its vertices.

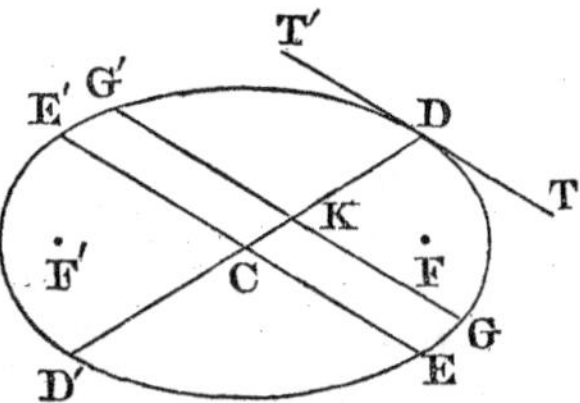

Thus, let DD′ be any diameter, and TT′ a tangent to the ellipse at D. From any point G of the curve draw GKG′ parallel to TT′ and cutting DD′ in K; then is GK an ordinate to the diameter DD′.

It is proved in Prop. XIX., Cor. 1, that GK is equal to G′K; hence the entire line GG′ is called a *double ordinate.*

11. The parts into which a diameter is divided by an ordinate, are called *abscissas.*

Thus, DK and D′K are the abscissas of the diameter DD′ corresponding to the ordinate GK.

12. Two diameters are *conjugate* to one another, when each is parallel to the ordinates of the other.

Thus, draw the diameter EE′ parallel to GK, an ordinate to the diameter DD′, in which case it will, of course, be parallel to the tangent TT′; then is the diameter EE′ conjugate to DD′.

13. The *latus rectum* is the double ordinate to the major axis which passes through one of the foci.

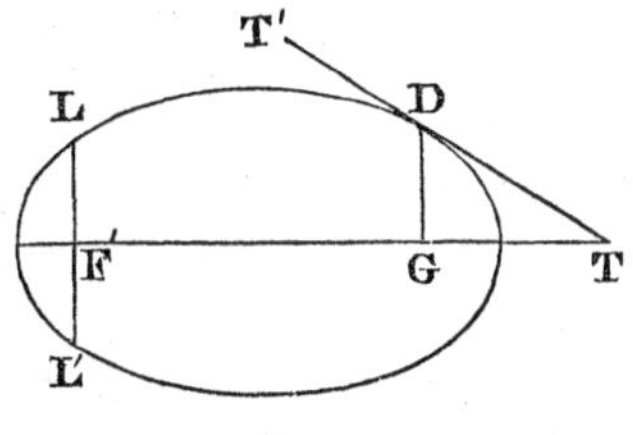

Thus, through the focus F′ draw LL′ a double ordinate to the major axis, it will be the latus rectum of the ellipse.

14. A *subtangent* is that part of the axis produced which is included between a tangent and the ordinate drawn from the point of contact.

Thus, if TT′ be a tangent to the curve at D, and DG an ordinate to the major axis, then GT is the corresponding subtangent.

PROPOSITION I. PROBLEM.

To describe an ellipse.

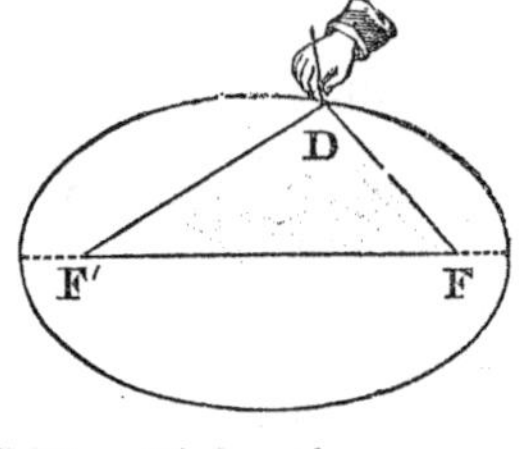

Let F and F′ be any two fixed points. Take a thread longer than the distance FF′, and fasten one of its extremities at F, the other at F′. Then let a pencil be made to glide along the thread so as to keep it always stretched; the curve described by the point of the pencil will be an ellipse. For, in every position of the pencil, the sum of the distances DF, DF′ will be the same, viz., equal to the entire length of the string.

PROPOSITION II. THEOREM.

The sum of the two lines drawn from any point of an ellipse to the foci, is equal to the major axis.

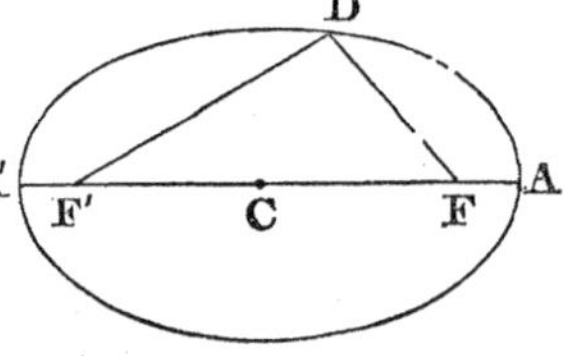

Let ADA′ be an ellipse, of which F, F′ are the foci, AA′ is the major axis, and D any point of the curve; then will DF+DF′ be equal to AA′.

For, by Def. 1, the sum of the distances of any point of the curve from the foci, is equal to a given line. Now, when the point D arrives at A, FA+F′A or 2AF+FF′ is equal to the given line. And when D is at A′, FA′+F′A′ or 2A′F′+FF′ is equal to the same line. Hence

$$2AF+FF'=2A'F'+FF';$$

consequently, AF is equal to A′F′.

Hence DF+DF′, which is equal to AF+AF′, must be equal to AA′. Therefore, the sum of the two lines, &c.

Cor. The major axis is bisected in the center. For, by Def. 3, CF is equal to CF′; and we have just proved that AF is equal to A′F′; therefore AC is equal to A′C.

PROPOSITION III. THEOREM.

Every diameter is bisected in the center.

Let D be any point of an ellipse; join DF, DF′, and FF′. Complete the parallelogram DFD′F′, and join DD′.

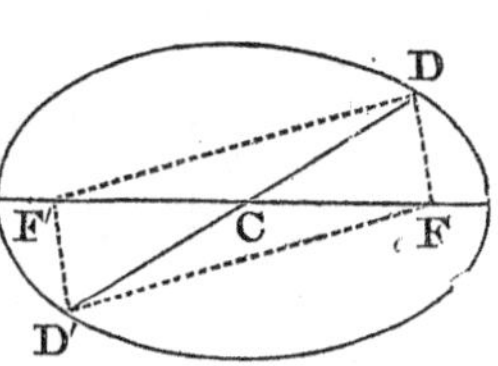

Now, because the opposite sides of a parallelogram are equal, the sum of DF and DF′ is equal to the sum of D′F and D′F′, hence D′ is a point in the ellipse. But the diagonals of a parallelogram bisect each other; therefore FF′ is bisected in C; that is, C is the center of the ellipse, and DD′ is a diameter bisected in C. Therefore, every diameter, &c.

PROPOSITION IV. THEOREM.

The distance from either focus to the extremity of the minor axis, is equal to half the major axis.

Let F and F′ be the foci of an ellipse, AA′ the major axis, and BB′ the minor axis; draw the straight lines BF, BF′; then BF, BF′ are each equal to AC.

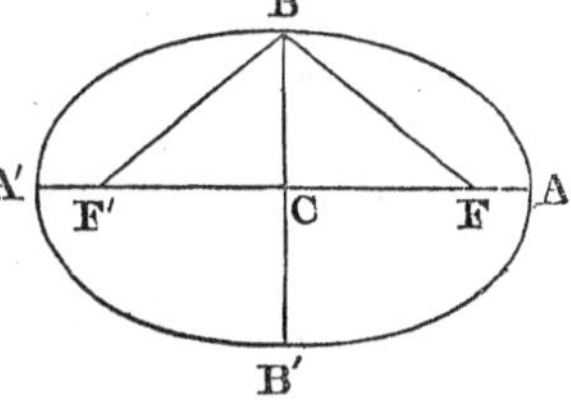

In the two right-angled triangles BCF, BCF′, CF is equal to CF′, and BC is common to both triangles; hence BF is equal to BF′. But BF+BF′ is equal to 2AC (Prop. II.); consequently, BF and BF′ are each equal to AC. Therefore, the distance, &c.

Cor. 1. *Half the minor axis is a mean proportional between the distances from either focus to the principal vertices.*

For BC^2 is equal to BF^2-FC^2 (Prop. XI., B. IV.), which is equal to AC^2-FC^2 (Prop. IV.). Hence (Prop. X., B. IV.),

$$BC^2=(AC+FC)\times(AC-FC)$$
$$=AF'\times AF\text{; and, therefore,}$$
$$AF : BC :: BC : FA'.$$

Cor. 2. *The square of the eccentricity is equal to the difference of the squares of the semi-axes.*

For FC^2 is equal to BF^2-BC^2, which is equal to AC^2-BC^2.

PROPOSITION V. THEOREM.

A tangent to the ellipse makes equal angles with straight lines drawn from the point of contact to the foci.

Let F, F′ be the foci of an ellipse, and D any point of the curve; if through the point D the line TT′ be drawn, making the angle TDF equal to T′DF′, then will TT′ be a tangent to the ellipse at D.

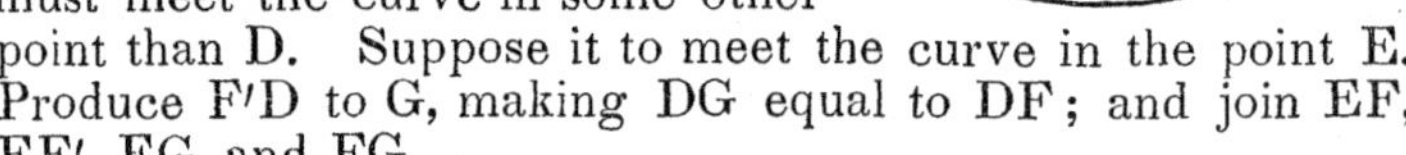

For if TT′ be not a tangent, it must meet the curve in some other point than D. Suppose it to meet the curve in the point E. Produce F′D to G, making DG equal to DF; and join EF, EF′, EG, and FG.

Now, in the two triangles DFH, DGH, because DF is equal to DG, DH is common to both triangles, and the angle FDH is, by supposition, equal to F′DT′, which is equal to the vertical angle GDH; therefore HF is equal to HG, and the angle DHF is equal to the angle DHG. Hence the line TT′ is perpendicular to FG at its middle point; and, therefore, EF is equal to EG.

Also, F′G is equal to F′D+DF, or F′E+EF, from the nature of the ellipse. But F′E+EG is greater than F′G (Prop. VIII., B. I.); it is, therefore, greater than F′E+EF. Consequently EG is greater than EF; which is impossible, for we have just proved EG equal to EF. Therefore E is not a point of the curve, and TT′ can not meet the curve in any other point than D; hence it is a tangent to the curve at the point D. Therefore, a tangent to the ellipse, &c.

Cor. 1. The tangents at the vertices of the axes, are perpendicular to the axes; and hence an ordinate to either axis is perpendicular to that axis.

Cor. 2. If TT′ represent a plane mirror, a ray of light proceeding from F in the direction FD, would be reflected in the direction DF′, making the angle of reflection equal to the angle of incidence. And, since the ellipse may be regarded as coinciding with a tangent at the point of contact, if rays of light proceed from one focus of a concave ellipsoidal mirror, they will all be reflected to the other focus. For this reason, the points F, F′ are called the *foci*, or burning points.

PROPOSITION VI. THEOREM.

Tangents to the ellipse at the vertices of a diameter, are parallel to each other.

Let DD be any diameter of an ellipse, and TT′ VV′ tangents to the curve at the points D, D′; then will they be parallel to each other.

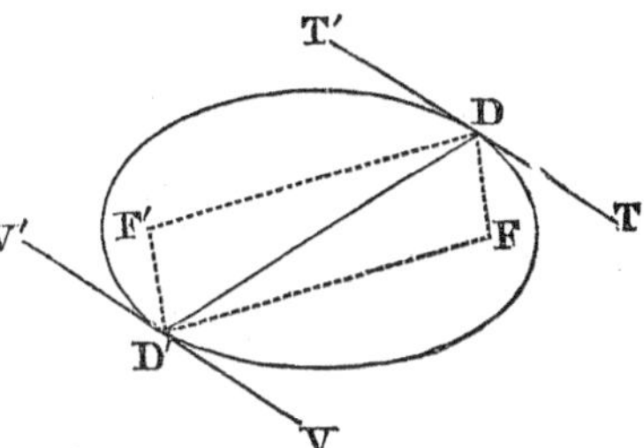

Join DF, DF′, D′F, D′F′; then, by the preceding Proposition, the angle FDT is equal to F′DT′, and the angle FD′V is equal to F′D′V′. But, by Prop. III., DFD′F′ is a parallelogram; and since the opposite angles of a parallelogram are equal, the angle FDF′ is equal to FD′F′; therefore the angle FDT is equal to F′D′V′ (Prop. II., B. I.). Also, since FD is parallel to F′D′, the angle FDD′ is equal to F′D′D; hence the whole angle D′DT is equal to DD′V′; and, consequently, TT′ is parallel to VV′. Therefore, tangents, &c.

Cor. If tangents are drawn through the vertices of any two diameters, they will form a parallelogram circumscribing the ellipse.

PROPOSITION VII. THEOREM.

If from the vertex of any diameter, straight lines are drawn through the foci, meeting the conjugate diameter, the part intercepted by the conjugate is equal to half the major axis.

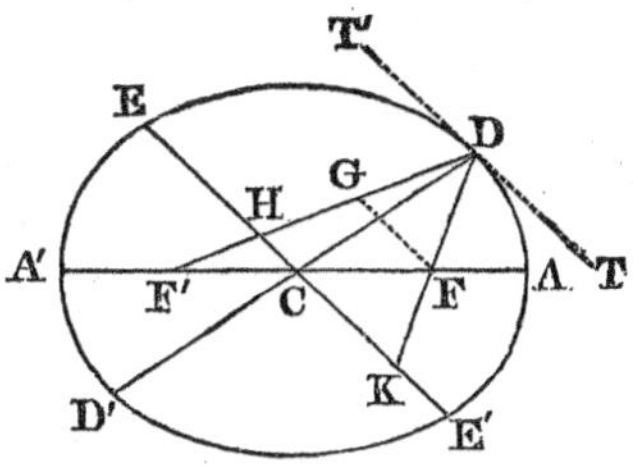

Let EE′ be a diameter conjugate to DD′, and let the lines DF, DF′ be drawn, and produced, if necessary, so as to meet EE′ in H and K; then will DH or DK be equal to AC.

Draw FG parallel to EE′ or TT′. Then the angle DGF is equal to the alternate angle F′DT′, and the angle DFG is equal to FDT. But the angles FDT, F′DT′ are equal to each other (Prop. V.); hence the angles DGF, DFG are equal to each other, and DG is equal to DF. Also, because CH is parallel to FG, and CF is equal to CF′; therefore HG must be equal to HF′.

Hence FD+F′D is equal to 2DG+2GH or 2DH. But FD+F′D is equal to 2AC. Therefore 2AC is equal to 2DH, or AC is equal to DH.

Also, the angle DHK is equal to DKH; and hence DK is equal to DH or AC. Therefore, if from the vertex, &c.

PROPOSITION VIII. THEOREM.

Perpendiculars drawn from the foci upon a tangent to the ellipse, meet the tangent in the circumference of a circle, whose diameter is the major axis.

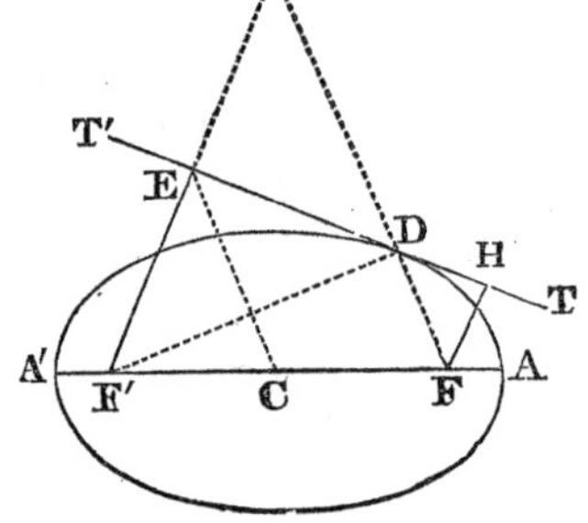

Let TT′ be a tangent to the ellipse at D, and from F′ draw F′E perpendicular to T′T; the point E will be in the circumference of a circle described upon AA′ as a diameter.

Join CE, FD, F′D, and produce F′E to meet FD produced in G.

Then, in the two triangles DEF′, DEG, because DE is common to both triangles, the angles at E are equal, being right angles; also, the angle EDF′ is equal to FDT (Prop. V.), which is equal to the vertical angle EDG; therefore DF′ is equal to DG, and EF′ is equal to EG.

Also, because F′E is equal to EG, and F′C is equal to CF, CE must be parallel to FG, and, consequently, equal to half of FG.

But, since DG has been proved equal to DF′, FG is equal to FD+DF′, which is equal to AA′. Hence CE is equal to half of AA′ or AC; and a circle described with C as a center, and radius CA, will pass through the point E. The same may be proved of a perpendicular let fall upon TT′ from the focus F. Therefore, perpendiculars, &c.

Cor. CE is parallel to DF, and if CH be joined, CH will be parallel to DF′.

PROPOSITION IX. THEOREM.

The product of the perpendiculars from the foci upon a tan gent, is equal to the square of half the minor axis.

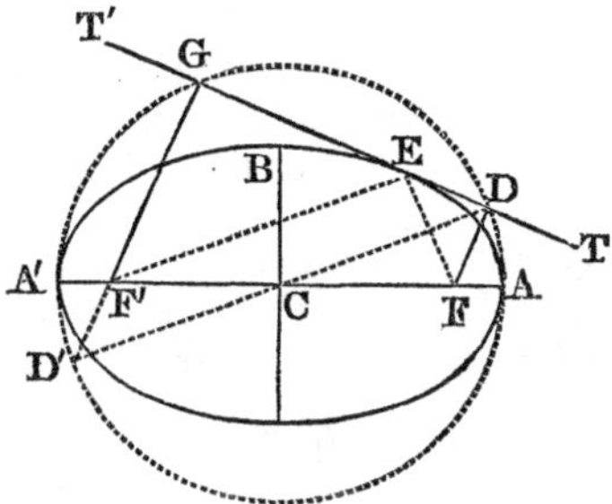

Let TT′ be a tangent to the ellipse at any point E, and let the perpendiculars FD, F′G be drawn from the foci; then will the product of FD by F′G, be equal to the square of BC.

On AA′, as a diameter, describe a circle; it will pass hrough the points D and G (Prop. VIII.). Join CD, and produce it to meet GF′ in D′. Then, because FD and F′G are perpendicular to the same straight line TT′, they are parallel to each other, and the alternate angles CFD, CF′D′ are equal. Also, the vertical angles DCF, D′CF′ are equal, and CF is equal to CF′. Therefore (Prop. VII., B. I.) DF is equal to D′F′, and CD is equal to CD′; that is, the point D′ is in the circumference of the circle ADGA′.

Hence DF×GF′ is equal to D′F′×GF′, which is equal to A′F′×F′A (Prop. XXVII., B. IV.), which is equal to BC^2 (Prop. IV., Cor. 1). Therefore, the product, &c.

Cor. The triangles FDE, F′GE are similar; hence

$$FD : F'G :: FE : F'E;$$

that is, *perpendiculars let fall from the foci upon a tangent, are to each other as the distances of the point of contact from the foci.*

PROPOSITION X. THEOREM.

If a tangent and ordinate be drawn from the same point of an ellipse, meeting either axis produced, half of that axis will be a mean proportional between the distances of the two inter sections from the center.

Let TT′ be a tangent to the ellipse, and DG an ordinate to the major axis from the point of contact; then we shall have

$$CT : CA :: CA : CG.$$

Join DF, DF′; then, since the exterior angle of the trian gle FDF′ is bisected by DT (Prop. V.), we have

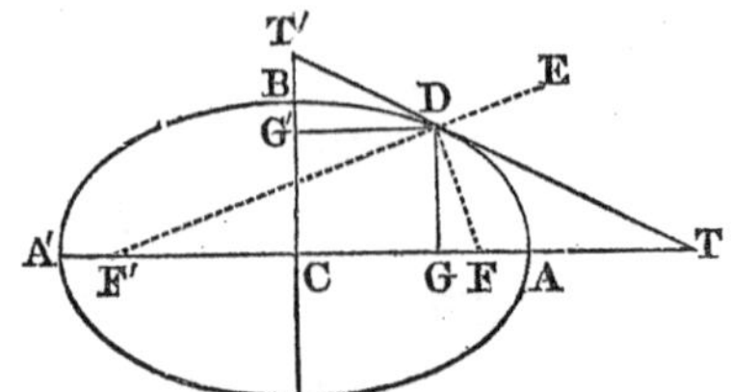

$F'T : FT :: F'D : FD$ (Prop. XVII., Sch., B. IV.).

Hence, by Prop. VII, Cor., B. II.,

$$F'T+FT : F'T-FT :: F'D+FD : F'D-FD,$$

or $$2CT : F'F :: 2CA : F'D-FD;$$

that is, $$2CT : 2CA :: F'F : F'D-FD. \quad (1)$$

Again, because DG is drawn from the vertex of the triangle FDF′ perpendicular to the base FF′, we have (Prop. XXXI., Cor., B. IV.),

$$F'F : F'D-FD :: F'D+FD : F'G-FG,$$

or $$F'F : F'D-FD :: 2CA : 2CG. \quad (2)$$

Comparing proportions (1) and (2), we have

$$2CT : 2CA :: 2CA : 2CG,$$

or $$CT : CA :: CA : CG.$$

It may also be proved that

$$CT' : CB :: CB : CG'.$$

Therefore, if a tangent, &c.

PROPOSITION XI. THEOREM.

The subtangent of an ellipse, is equal to the corresponding subtangent of the circle described upon its major axis.

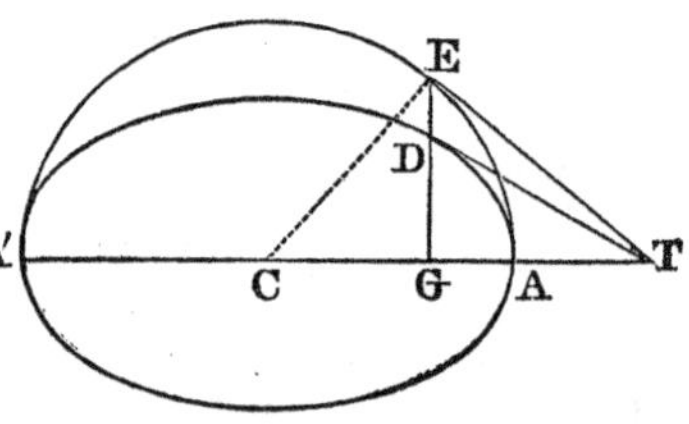

Let AEA′ be a circle described on AA′, the major axis of an ellipse; and from any point E in the circle, draw the ordinate EG cutting the ellipse in D. Draw DT touching the ellipse at D; join ET; then will ET be a tangent to the circle at E.

Join CE. Then, by the last Proposition,

$$CT : CA :: CA : CG;$$

or, because CA is equal to CE,

$$CT : CE :: CE : CG.$$

Hence the triangles CET, CGE, having the angle at C common, and the sides about this angle proportional, are similar Therefore the angle CET, being equal to the angle CGE, is a right angle; that is, the line ET is perpendicular to the radius CE, and is, consequently, a tangent to the circle (Prop. IX., B. III.). Hence GT is the subtangent corresponding to each of the tangents DT and ET. Therefore, the subtangent, &c.

Cor. A similar property may be proved of a tangent to the ellipse meeting the minor axis.

PROPOSITION XII. THEOREM.

The square of either axis, is to the square of the other, as the rectangle of the abscissas of the former, is to the square of their ordinate.

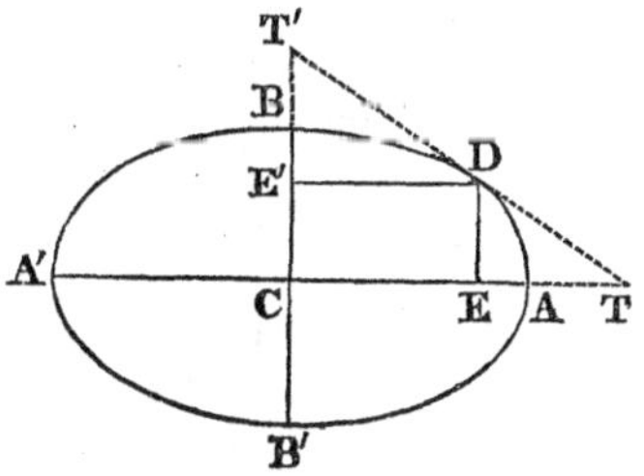

Let DE be an ordinate to the major axis from the point D; then we shall have

$CA^2 : CB^2 :: AE \times EA' : DE^2$.

Draw TT′ a tangent to the ellipse at D, then, by Prop. X.,

$CT : CA :: CA : CE$.

Hence (Prop. XII., B. II.),

$CA^2 : CE^2 :: CT : CE$;

and, by division (Prop. VII., B. II.),

$$CA^2 : CA^2 - CE^2 :: CT : ET. \qquad (1)$$

Again, by Prop. X.,

$CT' : CB :: CB : CE'$ or DE.

Hence (Prop. XII., B. II.),

$CB^2 : DE^2 :: CT' : DE$.

But, by similar triangles,

$CT' : DE :: CT : ET$;

therefore

$$CB^2 : DE^2 :: CT : ET. \qquad (2)$$

Comparing proportions (1) and (2), we have

$CA^2 : CA^2 - CE^2 :: CB^2 : DE^2$.

But $CA^2 - CE^2$ is equal to $AE \times EA'$ (Prop. X., B. IV.); hence

$CA^2 : CB^2 :: AE \times EA' : DE^2$.

In the same manner it may be proved that

$CB^2 : CA^2 :: BE' \times E'B' : DE'^2$.

Therefore, the square, &c.

Cor. 1. $CA^2 : CB^2 :: CA^2 - CE^2 : DE^2$.

Cor. 2. The squares of the ordinates to either axis, are to each other as the rectangles of their abscissas.

Cor. 3. *If a circle be described on either axis, then any ordinate in the circle, is to the corresponding ordinate in the ellipse, as the axis of that ordinate, is to the other axis.*

For, by the Proposition,

$CA^2 : CB^2 :: AE \times EA' : DE^2$.

But $AE \times EA'$ is equal to GE^2 (Prop. XXII., Cor., B. IV.).

Therefore $CA^2 : CB^2 :: GE^2 : DE^2$,

or $CA : CB :: GE : DE$.

In the same manner it may be proved that

$CB : CA :: G'E' : DE'$.

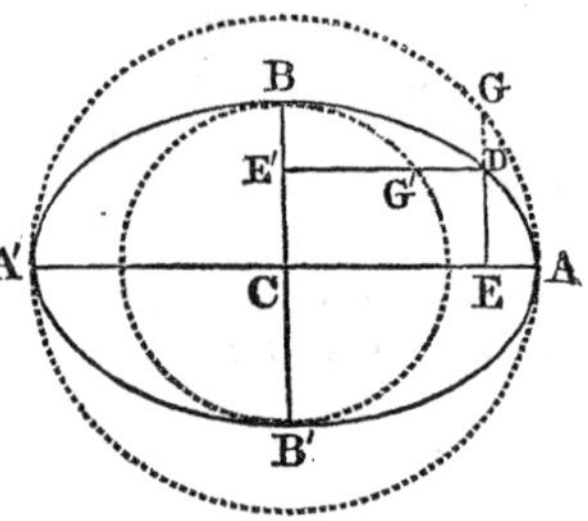

PROPOSITION XIII. THEOREM.

The latus rectum is a third proportional to the major and minor axes.

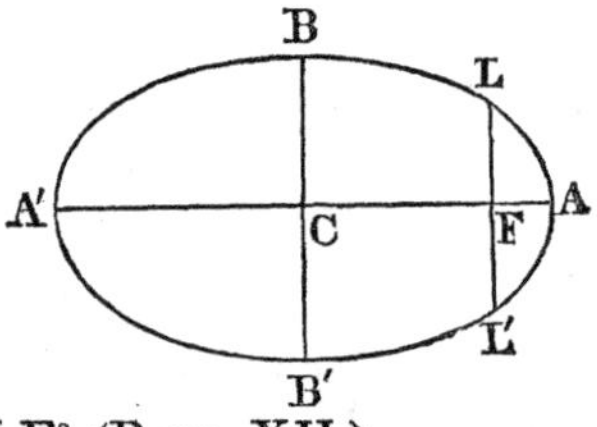

Let LL' be a double ordinate to the major axis passing through the focus F; then we shall have

$AA' : BB' :: BB' : LL'$.

Because LF is an ordinate to the major axis,

$AC^2 : BC^2 :: AF \times FA' : LF^2$ (Prop. XII.).

$:: BC^2 : LF^2$ (Prop. IV., Cor. 1).

Hence $AC : BC :: BC : LF$,

or $AA' : BB' :: BB' : LL'$.

Therefore, the latus rectum, &c.

PROPOSITION XIV. THEOREM.

If from the vertices of two conjugate diameters, ordinates are drawn to either axis, the sum of their squares will be equal to the square of half the other axis.

Let DD', EE' be any two conjugate diameters, DG and EH ordinates to the major axis drawn from their vertices; in which case, CG and CH will be equal to the ordinates to the minor axis drawn from the same points; then we shall have

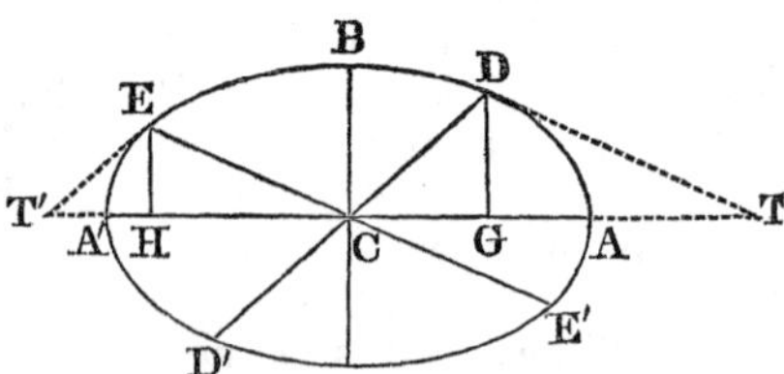

$$CA^2 = CG^2 + CH^2, \text{ and } CB^2 = DG^2 + EH^2.$$

Let DT be a tangent to the ellipse at D, and ET' a tangent at E. Then, by Prop. X.,

$CG \times CT$ is equal to CA^2, or $CH \times CT'$;

whence $CG : CH :: CT' : CT$; or, by similar triangles,

$:: CE : DT$; that is,

$:: CH : GT$.

Hence
$$CH^2 = GT \times CG,$$
$$= (CT - CG) \times CG$$
$$= CG \times CT - CG^2$$
$$= CA^2 - CG^2 \text{ (Prop. X.)};$$
that is,
$$CA^2 = CG^2 + CH^2.$$
In the same manner it may be proved that
$$CB^2 = DG^2 + EH^2.$$
Therefore, if from the vertices, &c.

Cor. 1. CH^2 is equal to $CA^2 - CG^2$; that is, $CG \times GT$; hence (Prop. XII., Cor. 1),
$$CA^2 : CB^2 :: CG \times GT : DG^2.$$

Cor. 2. CG^2 is equal to $CA^2 - CH^2$ or $AH \times HA'$; hence
$$CA^2 . CB^2 :: CG^2 : EH^2.$$

PROPOSITION XV. THEOREM.

The sum of the squares of any two conjugate diameters, is equal to the sum of the squares of the axes.

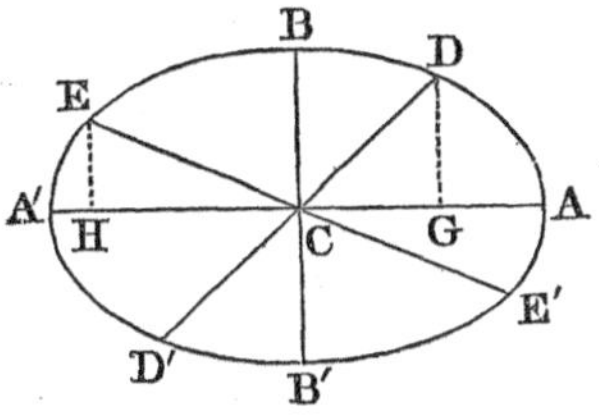

Let DD', EE' be any two conjugate diameters; then we shall have
$$DD'^2 + EE'^2 = AA'^2 + BB'^2.$$
Draw DG, EH ordinates to the major axis. Then, by the preceding Proposition,
$$CG^2 + CH^2 = CA^2,$$
and
$$DG^2 + EH^2 = CB^2.$$
Hence
$$CG^2 + DG^2 + CH^2 + EH^2 = CA^2 + CB^2,$$
or
$$CD^2 + CE^2 = CA^2 + CB^2;$$
that is,
$$DD'^2 + EE'^2 = AA'^2 + BB'^2.$$
Therefore, the sum of the squares, &c.

PROPOSITION XVI. THEOREM.

The parallelogram formed by drawing tangents through the vertices of two conjugate diameters, is equal to the rectangle of the axes.

Let $DED'E'$ be a parallelogram, formed by drawing tangents to the ellipse through the vertices of two conjugate diameters DD', EE'; its area is equal to $AA' \times BB'$.

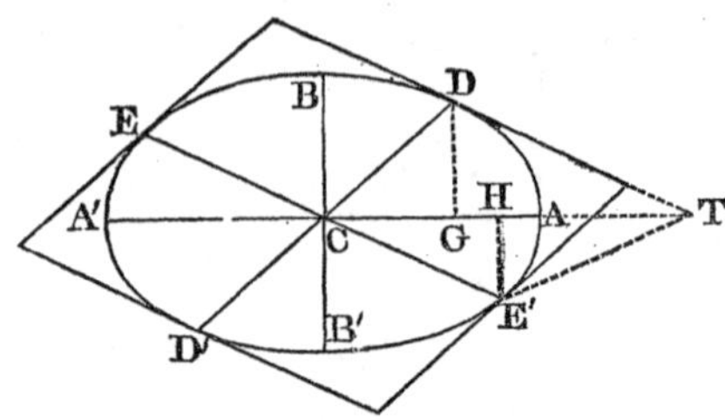

Let the tangent at D, meet the major axis produced in T, join E′T, and draw the ordinates DG, E′H.

Then, by Prop. XIV., Cor. 2, we have

$$CA^2 : CB^2 :: CG^2 : E'H^2,$$

or $$CA : CB :: CG : E'H.$$

But $$CT : CA :: CA : CG \text{ (Prop. X.)};$$

hence $$CT : CB :: CA : E'H,$$

or $$CA \times CB \text{ is equal to } CT \times E'H,$$

which is equal to twice the triangle CE′T, or the parallelogram DE′; since the triangle and parallelogram have the same base CE′, and are between the same parallels.

Hence $4CA \times CB$ or $AA' \times BB'$, is equal to $4DE'$, or the parallelogram DED′E′. Therefore, the parallelogram, &c.

PROPOSITION XVII. THEOREM.

If from the vertex of any diameter, straight lines are drawn to the foci, their product is equal to the square of half the conjugate diameter.

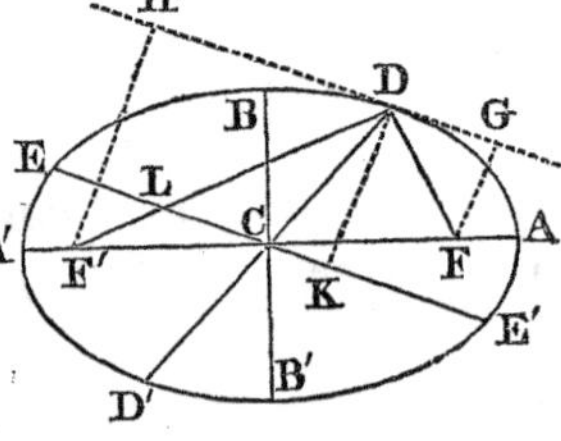

Let DD′, EE′ be two conjugate diameters, and from D let lines be drawn to the foci; then will $FD \times F'D$ be equal to EC^2.

Draw a tangent to the ellipse at D, and upon it let fall the perpendiculars FG, F′H; draw, also, DK perpendicular to EE′.

Then, because the triangles DFG, DLK, DF′H are similar, we have

$$FD : FG :: DL : DK.$$

Also, $$F'D : F'H :: DL : DK.$$

Whence (Prop. XI., B. II.),

$$FD \times F'D : FG \times F'H :: DL^2 : DK^2. \qquad (1)$$

But, by Prop. XVI $$AC \times BC = EC \times DK;$$

whence $$AC \text{ or } DL : DK :: EC : BC,$$

and $$DL^2 : DK^2 :: EC^2 : BC^2. \qquad (2)$$

Comparing proportions (1) and (2), we have

$$FD \times F'D : FG \times F'H :: EC^2 : BC^2.$$

But $FG \times F'H$ is equal to BC^2 (Prop. IX.); hence $FD \times F'D$ is equal to EC^2. Therefore, if from the vertex, &c.

PROPOSITION XVIII. THEOREM.

If a tangent and ordinate be drawn from the same point of an ellipse to any diameter, half of that diameter will be a mean proportional between the distances of the two intersections from the center.

Let a tangent EG and an ordinate EH be drawn from the same point E of an ellipse, meeting the diameter CD produced; then we shall have

$$CG : CD :: CD : CH.$$

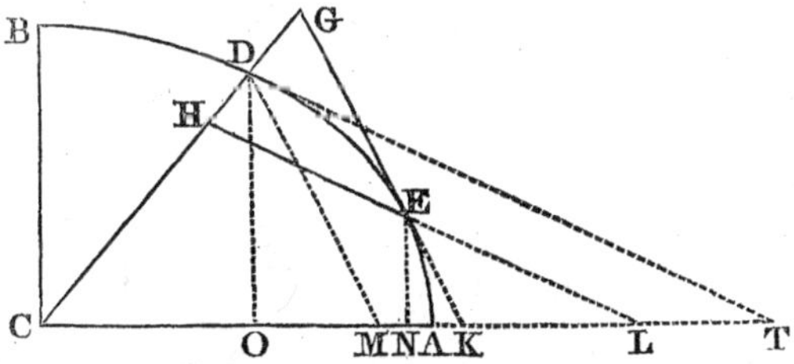

B G D H E C O MNAK L T

Produce EG and EH to meet the major axis in K and L; draw DT a tangent to the curve at the point D, and draw DM parallel to GK. Also, draw the ordinates EN, DO.

By Prop. XIV., Cor. 1, $CA^2 : CB^2 :: CO \times OT : DO^2,$
$$:: CN \times NK : EN^2.$$

Hence

$$CO \times OT : CN \times NK :: DO^2 : EN^2$$
$$:: OT^2 : NL^2, \text{ by similar triangles.} \quad (1)$$

Also, by similar triangles, $OT : NL :: DO : EN$
$$:: OM : NK. \quad (2)$$

Multiplying together proportions (1) and (2) (Prop. XI., B. II.), and omitting the factor OT^2 in the antecedents, and $NK \times NL$ in the consequents, we have

$$CO : CN :: OM : NL;$$

and, by composition, $CO : CN :: CM : CL.$ (3)

Also, by Prop. X., $CK \times CN = CA^2 = CT \times CO;$

hence $CO : CN :: CK : CT.$ (4)

Comparing proportions (3) and (4), we have

$$CK : CM :: CT : CL.$$

But $CK : CM :: CG : CD,$

and $CT : CL :: CD : CH;$

hence $CG : CD :: CD : CH.$

Therefore, if a tangent, &c.

PROPOSITION XIX. THEOREM.

The square of any diameter, is to the square of its conjugate, as the rectangle of its abscissas, is to the square of their ordinate.

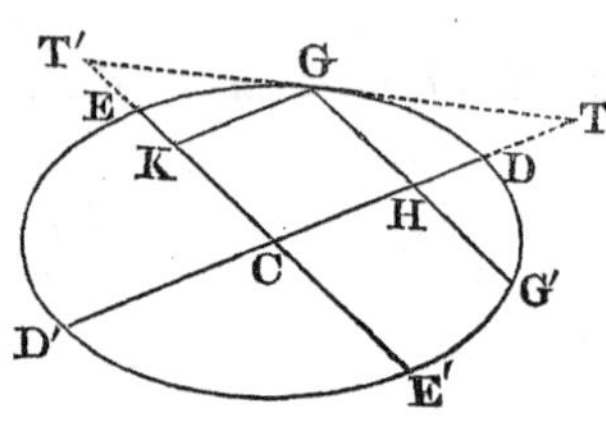

Let DD′, EE′ be two conjugate diameters, and GH an ordinate to DD′; then

$DD'^2 : EE'^2 :: DH \times HD' : GH^2.$

Draw TT′ a tangent to the curve at the point G, and draw GK an ordinate to EE′. Then, by Prop. XVIII.,

$$CT : CD :: CD : CH,$$

and $CD^2 : CH^2 :: CT : CH$ (Prop. XII., B. II.);

whence, by division,

$$CD^2 : CD^2 - CH^2 :: CT : HT. \qquad (1)$$

Also, by Prop. XVIII.,

$$CT' : CE :: CE : CK,$$

and $CE^2 : CK^2 :: CT' : CK$ or GH,

$$:: CT : HT. \qquad (2)$$

Comparing proportions (1) and (2), we have

$$CD^2 : CE^2 :: CD^2 - CH^2 : CK^2 \text{ or } GH^2,$$

or $DD'^2 : EE'^2 :: DH \times HD' : GH^2.$

Therefore, the square, &c.

Cor. 1. In the same manner it may be proved that $DD'^2 : EE'^2 :: DH \times HD' : G'H^2$; hence GH is equal to G′H, or *every diameter bisects its double ordinates.*

Cor. 2. The squares of the ordinates to any diameter, are to each other as the rectangles of their abscissas.

PROPOSITION XX. THEOREM.

If a cone be cut by a plane, making an angle with the base less than that made by the side of the cone, the section is an ellipse.

Let ABC be a cone cut by a plane DEGH, making an angle with the base, less than that made by the side of the cone; the section D*e*EGH*h* is an ellipse.

Let ABC be a section through the axis of the cone, and perpendicular to the plane DEGH. Let EMHO, *emho* be circular sections parallel to the base; then EH, the intersection of the planes DEGH, EMHO, will be perpendicular to the plane ABC, and, consequently, to each of the lines DG, MO. So, also, eh will be perpendicular to DG and mo.

Now, because the triangles DNO, Dno are similar, as also the triangles GMN, Gmn, we have the proportions,

$$NO : no :: DN : Dn,$$

and $$MN : mn :: NG : nG.$$

Hence, by Prop. XI., B. II.,

$$MN \times NO : mn \times no :: DN \times NG : Dn \times nG.$$

But since MO is a diameter of the circle EMHO, and EN is perpendicular to MO, we have (Prop. XXII., Cor., B. IV.).

$$MN \times NO = EN^2.$$

For the same reason, $mn \times no = en^2$.

Substituting these values of $MN \times NO$ and $mn \times no$, in the preceding proportion, we have

$$EN^2 : en^2 :: DN \times NG : Dn \times nG;$$

that is, the squares of the ordinates to the diameter DG, are to each other as the products of the corresponding abscissas. Therefore the curve is an ellipse (Prop. XII., Cor. 2) whose major axis is DG. Hence the ellipse is called a *conic section*, as mentioned on page 177.

PROPOSITION XXI. THEOREM.

The area of an ellipse is a mean proportional between the two circles described on its axes.

Let AA′ be the major axis of an ellipse ABA′B′. On AA′ as a diameter, describe a circle; inscribe in the circle any regular polygon AEDA′, and from the vertices E, D, &c., of the polygon, draw perpendiculars to AA′. Join the points B, G, &c., in which these perpendiculars intersect the ellipse, and there will be inscribed in the ellipse a polygon of an equal number of sides.

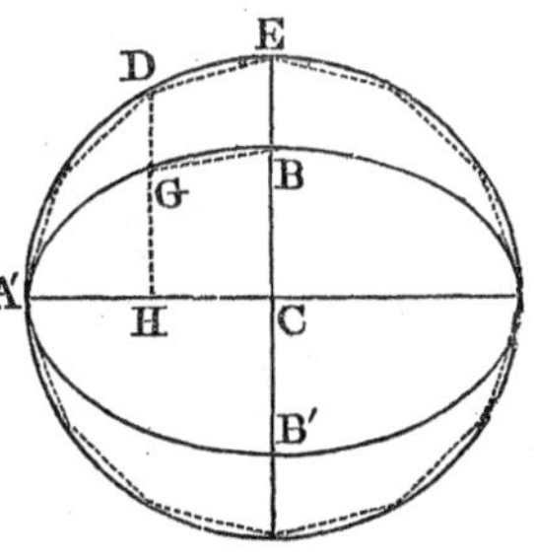

Now the area of the trapezoid CEDH, is equal to $(CE + DH) \times \frac{CH}{2}$; and the area of the trapezoid CBGH, is equal to

$(CB+GH)\times\frac{CH}{2}$. These trapezoids are to each other, as CE+DH to CB+GH, or as AC to BC (Prop. XII., Cor. 3).

In the same manner it may be proved that each of the trapezoids composing the polygon inscribed in the circle, is to the corresponding trapezoid of the polygon inscribed in the ellipse, as AC to BC. Hence, the entire polygon inscribed in the circle, is to the polygon inscribed in the ellipse, as AC to BC.

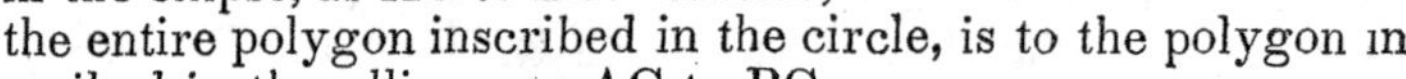

Since this proportion is true, whatever be the number of sides of the polygons, it will be true when the number is indefinitely increased; in which case one of the polygons coincides with the circle, and the other with the ellipse. Hence we have

Area of circle : *area of ellipse* : : AC : BC.

But the area of the circle is represented by πAC^2; hence the area of the ellipse is equal to $\pi AC \times BC$, which is a mean proportional between the two circles described on the axes.

HYPERBOLA.

Definitions.

1. An *hyperbola* is a plane curve, in which the difference of the distances of each point from two fixed points, is equal to a given line.

2. The two fixed points are called the *foci*.

Thus, if F and F′ are two fixed points, and if the point D moves about F in such a manner that the *difference* of its distances from F and F′ is always the same, the point D will describe an hyperbola, of which F and F′ are the foci.

If the point D′ moves about F′ in such a manner that D′F—D′F′ is always equal to DF′—DF, the point D′ will describe a second hyperbola similar to the first. The two curves are called *opposite hyperbolas.*

3. The *center* is the middle point of the straight line joining the foci.

4. The *eccentricity* is the distance from the center to either focus.

Thus, let F and F′ be the foci of two opposite hyperbolas. Draw the line FF′, and bisect it in C. The point C is the center of the hyperbola, and CF or CF′ is the eccentricity.

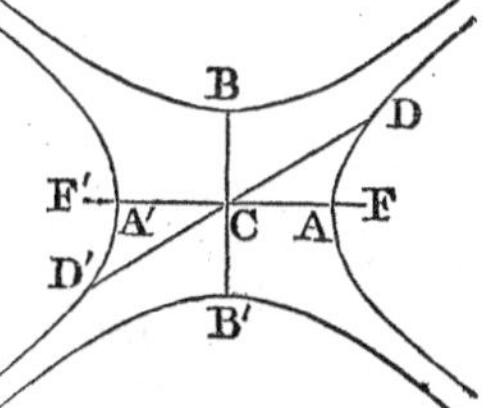

5. A *diameter* is a straight line drawn through the center, and terminated by two opposite hyperbolas.

6. The extremities of a diameter are called its *vertices.*

Thus, through C draw any straight line DD′ terminated by the opposite curves; DD′ is a diameter of the hyperbola; D and D′ are its vertices.

7. The *major axis* is the diameter which, when produced, passes through the foci; and its extremities are called the *principal vertices.*

8. The *minor axis* is a line drawn through the center per-

pendicular to the major axis, and terminated by the circumference described from one of the principal vertices as a center, and a radius equal to the eccentricity.

Thus, through C draw BB′ perpendicular to AA′, and with A as a center, and with CF as a radius, describe a circumference cutting this perpendicular in B and B′; then AA′ is the major axis, and BB′ the minor axis.

If on BB′ as a major axis, opposite hyperbolas are described, having AA′ as their minor axis, these hyperbolas are said to be *conjugate* to the former.

9. A *tangent* is a straight line which meets the curve, but, being produced, does not cut it.

10. An *ordinate* to a diameter, is a straight line drawn from any point of the curve to meet the diameter produced, parallel to the tangent at one of its vertices.

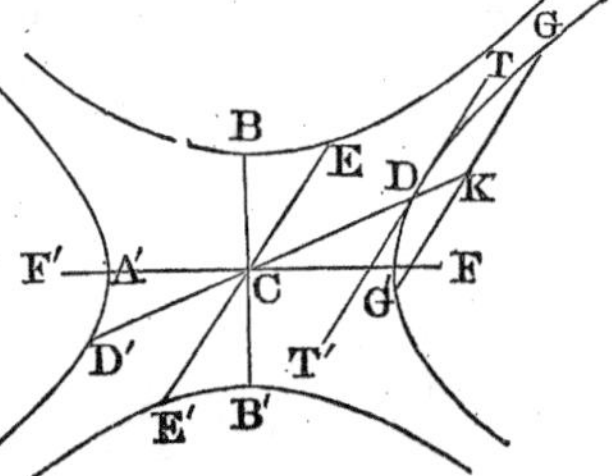

Thus, let DD′ be any diameter, and TT′ a tangent to the hyperbola at D. From any point G of the curve draw GKG′ parallel to TT′ and cutting DD′ produced in K; then is GK an ordinate to the diameter DD′.

It is proved, in Prop. XIX., Cor. 1, that GK is equal to G′K; hence the entire line GG′ is called a *double ordinate.*

11. The parts of the diameter produced, intercepted between its vertices and an ordinate, are called its *abscissas.*

Thus, DK and D′K are the abscissas of the diameter DD′ corresponding to the ordinate GK.

12. Two diameters are *conjugate* to one another, when each is parallel to the ordinates of the other.

Thus, draw the diameter EE′ parallel to GK an ordinate to the diameter DD′, in which case it will, of course, be parallel to the tangent TT′; then is the diameter EE′ conjugate to DD′.

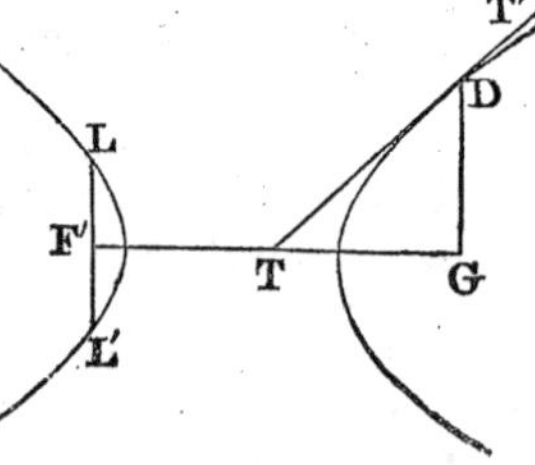

13. The *latus rectum* is the double ordinate to the major axis which passes through one of the foci.

Thus, through the focus F′ draw LL′ a double ordinate to the major axis, it will be the latus rectum of the hyperbola.

15. A *subtangent* is that part of the axis produced which is included between a tangent, and the ordinate drawn from the point of contact.

Thus, if TT′ be a tangent to the curve at D, and DG an ordinate to the major axis, then GT is the corresponding subtangent.

PROPOSITION I. PROBLEM.

To describe an hyperbola.

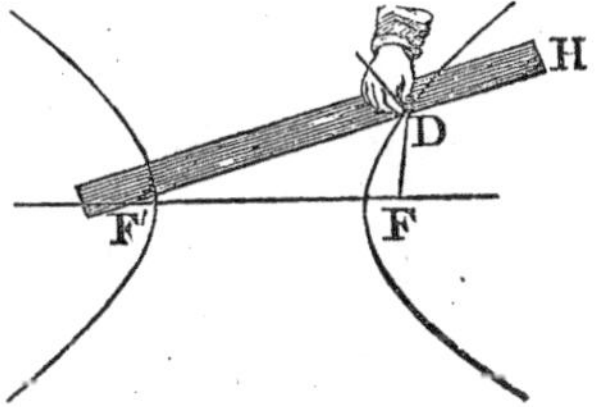

Let F and F′ be any two fixed points. Take a ruler longer than the distance FF′, and fasten one of its extremities at the point F′. Take a thread shorter than the ruler, and fasten one end of it at F, and the other to the end H of the ruler. Then move the ruler HDF′ about the point F′, while the thread is kept constantly stretched by a pencil pressed against the ruler; the curve described by the point of the pencil, will be a portion of an hyperbola. For, in every position of the ruler, the difference of the lines DF, DF′ will be the same, viz., the difference between the length of the ruler and the length of the string.

If the ruler be turned, and move on the other side of the point F, the other part of the same hyperbola may be described. Also, if one end of the ruler be fixed in F, and that of the thread in F′, the opposite hyperbola may be described.

PROPOSITION II. THEOREM.

The difference of the two lines drawn from any point of an hyperbola to the foci, is equal to the major axis.

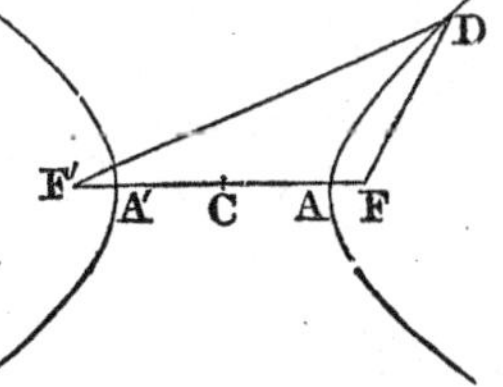

Let F and F′ be the foci of two opposite hyperbolas, AA′ the major axis, and D any point of the curve; then will DF′−DF be equal to AA′.

For, by Def. 1, the difference of the distances of any point of the curve from the foci, is equal to a given line. Now when the point D arrives at A, F′A−FA, or AA′+F′A′−FA, is equal to the given line. And when D is at A′, FA′−F′A′, or AA′+AF−A′F′, is equal to the same line. Hence

$$AA'+AF-A'F'=AA'+F'A'-FA,$$

or
$$2AF=2A'F';$$

that is, AF is equal to A′F′.

Hence DF′−DF, which is equal t AF′−AF, must be equal to AA′. Therefore, the difference of the two lines, &c.

Cor. The major axis is bisected in the center. For, by Def. 3, CF is equal to CF′; and we have just proved that AF is equal to A′F′; therefore AC is equal to A′C.

PROPOSITION III. THEOREM.

Every diameter is bisected in the center.

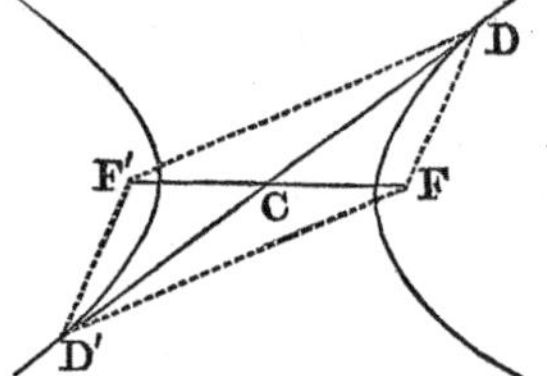

Let D be any point of an hyperbola; join DF, DF′, and FF′. Complete the parallelogram DFD′F′, and join DD′.

Now, because the opposite sides of a parallelogram are equal, the difference between DF and DF′ is equal to the difference between D′F and D′F′; hence D′ is a point in the opposite hyperbola. But the diagonals of a parallelogram bisect each other; therefore FF′ is bisected in C; that is, C is the center of the hyperbola, and DD′ is a diameter bisected in C. Therefore, every diameter, &c.

PROPOSITION IV. THEOREM.

Half the minor axis is a mean proportional between the distances from either focus to the principal vertices.

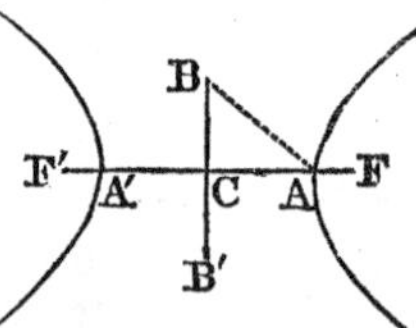

Let F and F′ be the foci of opposite hyperbolas, AA′ the major axis, and BB′ the minor axis; then will BC be a mean proportional between AF and A F.

Join AB. Now BC^2 is equal to AB^2-AC^2, which is equal to FC^2-AC^2 (Def. 8). Hence (Prop. X., B. IV.),

$$BC^2=(FC-AC)\times(FC+AC)$$
$$=AF\times A'F;$$

and hence
$$AF : BC :: BC : A'F.$$

Cor. The square of the eccentricity is equal to the sum of the squares of the semi-axes.

For FC^2 is equal to AB^2 (Def. 8), which is equal to AC^2+BC^2.

PROPOSITION V. THEOREM.

A tangent to the hyperbola bisects the angle contained by lines drawn from the point of contact to the foci.

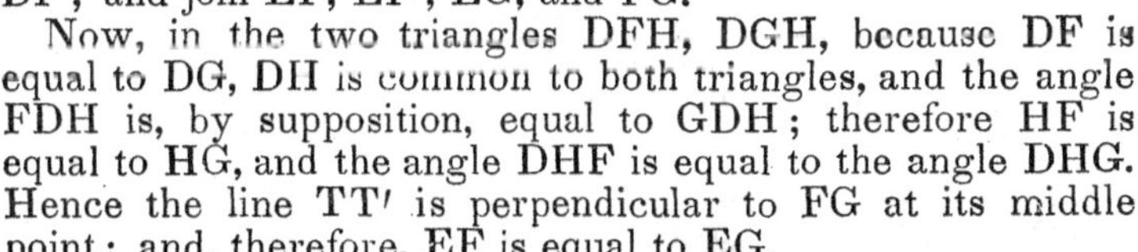

Let F, F′ be the foci of two opposite hyperbolas, and D any point of the curve; if through the point D, the line TT′ be drawn bisecting the angle FDF′; then will TT′ be a tangent to the hyperbola at D.

For if TT′ be not a tangent, let it meet the curve in some other point, as E. Take DG equal to DF; and join EF, EF′, EG, and FG.

Now, in the two triangles DFH, DGH, because DF is equal to DG, DH is common to both triangles, and the angle FDH is, by supposition, equal to GDH; therefore HF is equal to HG, and the angle DHF is equal to the angle DHG. Hence the line TT′ is perpendicular to FG at its middle point; and, therefore, EF is equal to EG.

Now F′G is equal to F′D − DF, or F′E − EF, from the nature of the hyperbola. But F′E − EG is less than F′G (Prop. VIII., B. I.); it is, therefore, less than F′E − EF. Consequently, EG is greater than EF, which is impossible, for we have just proved EG equal to EF. Therefore E is not a point of the curve; and TT′ can not meet the curve in any other point than D; hence it is a tangent to the curve at the point D. Therefore, a tangent to the hyperbola, &c.

Cor. 1. The tangents at the vertices of the axes, are perpendicular to the axes; and hence an ordinate to either axis is perpendicular to that axis.

Cor. 2. If TT′ represent a plane mirror, a ray of light proceeding from F in the direction FD, would be reflected in a line which, if produced, would pass through F′, making the angle of reflection equal to the angle of incidence. And, since the hyperbola may be regarded as coinciding with a tangent at the point of contact, if rays of light proceed from one focus of a concave hyperbolic mirror, they will be reflected in lines diverging from the other focus. For this reason, the points F, F′ are called the *foci*.

PROPOSITION VI. THEOREM.

Tangents to the hyperbola at the vertices of a diameter, are parallel to each other.

Let DD′ be any diameter of an hyperbola, and TT′, VV′ tangents to the curve at the points D, D′; then will they be parallel to each other.

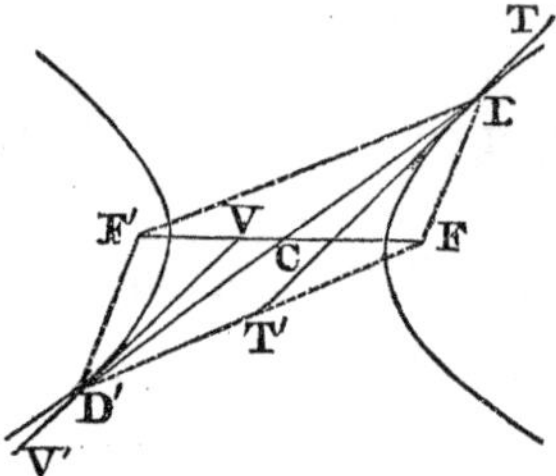

Join DF, DF′, D′F, D′F′. Then, by Prop. III., FDF′D′ is a parallelogram; and, since the opposite angles of a parallelogram are equal, the angle FDF′ is equal to FD′F′. But the tangents TT′, VV′ bisect the angles at D and D′ (Prop. V.); hence the angle F′DT′, or its alternate angle FT′D, is equal to FD′V. But FT′D is the exterior angle opposite to FD′V; hence TT′ is parallel to VV′. Therefore tangents, &c.

Cor. If tangents are drawn through the vertices of any two diameters, they will form a parallelogram.

PROPOSITION VII. THEOREM.

If through the vertex of any diameter, straight lines are drawn from the foci, meeting the conjugate diameter, the part intercepted by the conjugate is equal to half of the major axis.

Let EE′ be a diameter conjugate to DD′, and let the lines DF, DF′ be drawn, and produced, if necessary, so as to meet EE′ in H and K; then will DH or DK be equal to AC.

Draw F′G parallel to EE′ or TT′, meeting FD produced in G. Then the angle DGF′ is equal to the exterior angle FDT′; and the angle DF′G is equal to the alternate angle F′DT′. But the angles FDT′, F′DT′ are equal to each other (Prop. V.); hence the angles DGF′, DF′G are equal to each other, and DG is equal to DF′. Also, because CK is parallel to F′G, and CF is equal to CF′; therefore FK must be equal to KG.

Hence F′D—FD is equal to GD—FD or GF—2DF; that is, 2KF—2DF or 2DK. But F′D—FD is equal to 2AC. Therefore 2AC is equal to 2DK, or AC is equal to DK.

Also, the angle DHK is equal to DKH; and hence DH is equal to DK or AC. Therefore, if through the vertex, &c.

PROPOSITION VIII. THEOREM.

Perpendiculars drawn from the foci upon a tangent to the hyperbola, meet the tangent in the circumference of a circle whose diameter is the major axis.

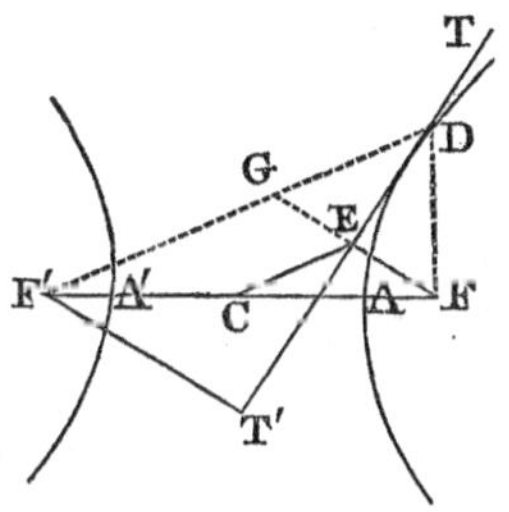

Let TT′ be a tangent to the hyperbola at D, and from F draw FE perpendicular to TT′; the point E will be in the circumference of a circle described upon AA′ as a diameter.

Join CE, FD, F′D, and produce FE to meet F′D in G.

Then, in the two triangles DEF, DEG, because DE is common to both triangles, the angles at E are equal, being right angles; also, the angle EDF is equal to EDG (Prop. V.); therefore DF is equal to DG, and EF to EG.

Also, because FE is equal to EG, and CF is equal to CF′, CE must be parallel to F′G, and, consequently, equal to half of F′G.

But, since DG has been proved equal to DF, F′G is equal to F′D—FD, which is equal to AA′. Hence CE is equal to half of AA′ or AC; and a circle described with C as a center, and radius CA, will pass through the point E. The same may be proved of a perpendicular let fall upon TT′ from the focus F′. Therefore, perpendiculars, &c.

PROPOSITION IX. THEOREM.

The product of the perpendiculars from the foci upon a tangent, is equal to the square of half the minor axis.

Let TT′ be a tangent to the hyperbola at any point E, and let the perpendiculars FD, F′G be drawn from the foci; then will the product of FD by F′G, be equal to the square of BC.

On AA′ as a diameter, describe a circle; it will pass through the points D and G (Prop. VIII.). Join CD, and produce it to meet GF′ in D′. Then, because FD and F′G are perpendicular to the same straight line TT′, they are parallel to each other, and the alternate angles CFD, CF′D′ are equal. Also, the vertical angles DCF, D′CF′ are equal, and CF is equal to CF′. Therefore (Prop. VII., B. I.), DF is equal to D′F′, and CD is equal to CD′; that is, the point D′ is in the circumference of the circle ADA′G.

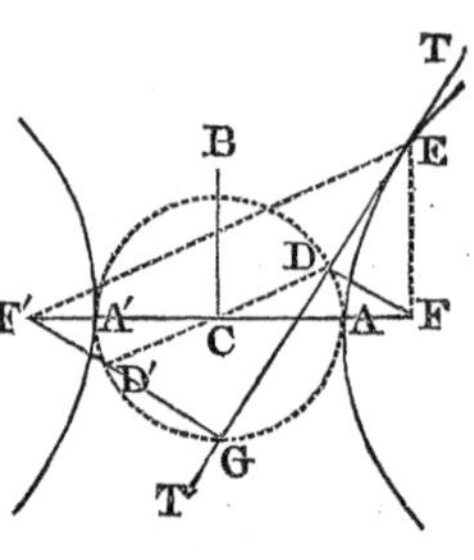

Hence DF×GF′ is equal to D′F′×GF′, which is equal to A′F′×F′A (Prop. XXVIII., Cor. 2, B. IV.), which is equal to BC^2 (Prop. IV.). Therefore, the product, &c.

Cor. The triangles FDE, F′GE are similar; hence

FD : F′G : : FE : F′E;

that is, *perpendiculars let fall from the foci upon a tangent, are to each other as the distances of the point of contact from the foci.*

PROPOSITION X. THEOREM.

If a tangent and ordinate be drawn from the same point of an hyperbola, meeting either axis produced, half of that axis will be a mean proportional between the distances of the two intersections from the center.

Let DTT′ be a tangent to the hyperbola, and DG an ordinate to the major axis from the point of contact; then we shall have

CT : CA : : CA : CG.

Join DF, DF′; then, since the angle FDF′ is bisected by DT (Prop. V.), we have

F′T : FT : : F′D : FD

(Prop. XVII., B. IV.).

Hence, by Prop. VII., Cor., B. II.,

F′T−FT : F′T+FT : : F′D−FD : F′D+FD,

or 2CT : F′F : : 2CA : F′D+FD;

that is, 2CT : 2CA : : F′F : F′D+FD. (1)

Again, because DG is drawn from the vertex of the triangle FDF′ perpendicular to the base FF′ produced, we have (Prop. XXXI., Cor., B. IV.),

F′F . F′D+FD : . F′D−FD : F′G+FG,

or F′F : F′D+FD : : 2CA : 2CG. (2)

Comparing proportions (1) and (2), we have

$$2CT : 2CA :: 2CA : 2CG,$$

or

$$CT : CA :: CA : CG.$$

It may also be proved that

$$CT' : CB :: CB : CG'.$$

Therefore, if a tangent, &c.

PROPOSITION XI. THEOREM.

The subtangent of an hyperbola, is equal to the corresponding subtangent of the circle described upon its major axis.

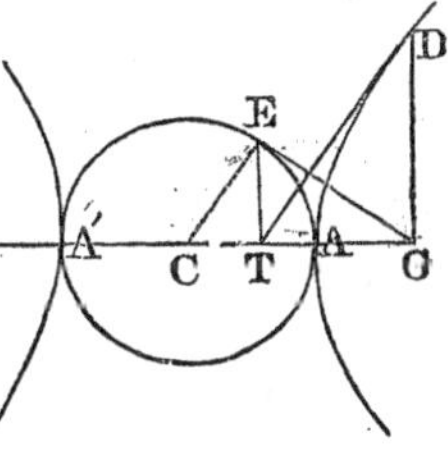

Let AEA′ be a circle described on AA′ the major axis of an hyperbola; and from any point E in the circle, draw the ordinate ET. Through T draw the line DT touching the hyperbola in D, and from the point of contact draw the ordinate DG. Join GE; then will GE be a tangent to the circle at E.

Join CE. Then, by the last Proposition,

$$CT : CA :: CA : CG;$$

or, because CA is equal to CE,

$$CT : CE :: CE : CG.$$

Hence the triangles CET, CGE having the angle at C common, and the sides about this angle proportional, are similar. Therefore the angle CEG, being equal to the angle CTE, is a right angle; that is, the line GE is perpendicular to the radius CE, and is, consequently, a tangent to the circle (Prop. IX., B. III.). Hence GT is the subtangent corresponding to each of the tangents DT and EG. Therefore, the subtangent, &c.

PROPOSITION XII. THEOREM.

The square of either axis, is to the square of the other, as the rectangle of the abscissas of the former, is to the square of their ordinate.

Let DE be an ordinate to the major axis from the point D; then we shall have

$$CA^2 : CB^2 :: AE \times EA' : DE^2.$$

Draw DTT′ a tangent to the hyperbola at D; then, by Prop. X.,

$$CT : CA :: CA : CE.$$

Hence (Prop. XII., B. II.)

$$CA^2 : CE^2 :: CT : CE;$$

and, by division (Prop. VII., B. II.),

$$CA^2 : CE^2 - CA^2 :: CT : ET. \quad (1)$$

Again, by Prop. X.,

$$CT' : CB :: CB : CE' \text{ or } DE.$$

Hence (Prop. XII., B. II.),

$$CB^2 : DE^2 :: CT' : DE.$$

But, by similar triangles,

$$CT' : DE :: CT : ET;$$

therefore $$CB^2 : DE^2 :: CT : ET. \quad (2)$$

Comparing proportions (1) and (2), we have

$$CA^2 : CE^2 - CA^2 :: CB^2 : DE^2.$$

But $CE^2 - CA^2$ is equal to $AE \times EA'$ (Prop. X., B. IV.); hence

$$CA^2 : CB^2 :: AE \times EA' : DE^2.$$

In the same manner it may proved that

$$CB^2 : CA^2 :: BE' \times E'B' : D'E'^2.$$

Therefore, the square, &c.

Cor. 1. $CA^2 : CB^2 :: CE^2 - CA^2 : DE^2$.

Cor. 2. The squares of the ordinates to either axis, are to each other as the rectangles of their abscissas.

Cor. 3. If a circle be described on the major axis, then any tangent to the circle, is to the corresponding ordinate in the hyperbola, as the major axis is to the minor axis.

For, by the Proposition,

$$CA^2 : CB^2 :: AE \times EA' : DE^2.$$

But $AE \times EA'$ is equal to GE^2 (Prop. XXVIII., B. IV.).

Therefore $$CA^2 : CB^2 :: GE^2 : DE^2,$$

or $$CA : CB :: GE : DE.$$

PROPOSITION XIII. THEOREM.

The latus rectum is a third proportional to the major and minor axes.

Let LL′ be a double ordinate to the major axis passing through the focus F; then we shall have

$$AA' : BB' :: BB' : LL'.$$

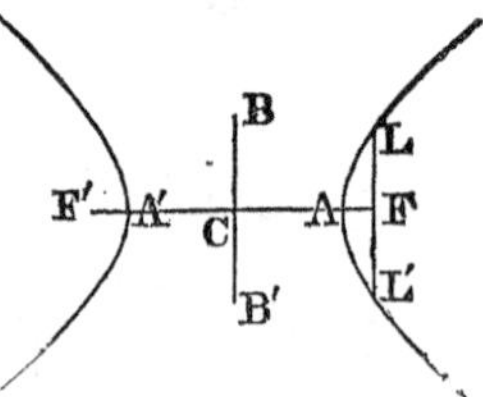

Because LF is an ordinate to the major axis,

$$AC^2 : BC^2 :: AF \times FA' : LF^2 \text{ (Prop. XII.)}$$

$$:: BC^2 : LF^2 \text{ (Prop. IV.)}$$

Hence $AC : BC :: BC : LF$,
or $AA' : BB' :: BB' : LL'$.
Therefore, the latus rectum, &c.

PROPOSITION XIV. THEOREM.

If from the vertices of two conjugate diameters, ordinates are drawn to either axis, the difference of their squares will be equal to the square of half the other axis.

Let DD′, EE′ be any two conjugate diameters, DG and EH ordinates to the major axis drawn from their vertices, in which case, CG and CH will be equal to the ordinates to the minor axis drawn from the same points; then we shall have $CA^2=CG^2-CH^2$, and $CB^2=EH^2-DG^2$.

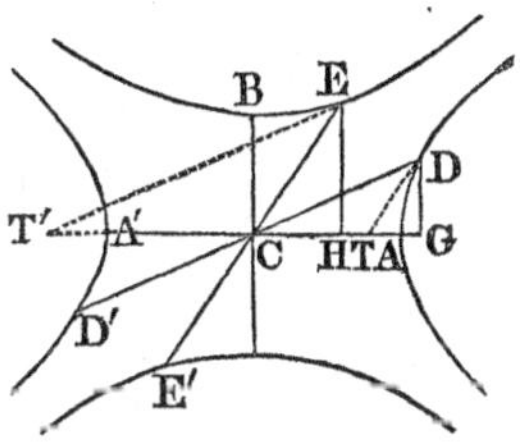

Let DT be a tangent to the curve at D, and ET′ a tangent at E. Then, by Prop. X.,

$CG \times CT$ is equal to CA^2, or $CH \times CT'$;

whence

$CG : CH :: CT' : CT$; or, by similar triangles,
$:: CE : DT$; that is,
$:: CH : GT$.

Hence $CH^2=GT \times CG$
$=(CG-CT) \times CG$
$=CG^2-CG \times CT$
$=CG^2-CA^2$ (Prop. X.);

that is $CA^2=CG^2-CH^2$.

In the same manner it may be proved that

$$CB^2=EH^2-DG^2.$$

Therefore, if from the vertices, &c.

Cor. 1. CH^2 is equal to CG^2-CA^2; that is, $CG \times GT$; hence (Prop. XII., Cor. 1),

$$CA^2 : CB^2 :: CG \times GT : DG^2.$$

Cor. 2. By Prop. XII.,

$$CB^2 : CA^2 :: EH^2-CB^2 : CH^2.$$

By composition,

$$CB^2 : CA^2 :: EH^2 : CA^2+CH^2 \text{ or } CG^2.$$

Hence $CA^2 : CB^2 :: CG^2 : EH^2$.

PROPOSITION XV. THEOREM.

The difference of the squares of any two conjugate diameters, is equal to the difference of the squares of the axes.

Let DD′, EE′ be any two conjugate diameters; then we shall have

$$DD'^2 - EE'^2 = AA'^2 - BB'^2.$$

Draw DG, EH ordinates to the major axis. Then, by the preceding Proposition,

$$CG^2 - CH^2 = CA^2,$$

and $\quad EH^2 - DG^2 = CB^2.$

Hence $\quad CG^2 + DG^2 - CH^2 - EH^2 = CA^2 - CB^2,$

or $\quad CD^2 - CE^2 = CA^2 - CB^2;$

that is, $\quad DD'^2 - EE'^2 = AA'^2 - BB'^2.$

Therefore, the difference of the squares, &c.

PROPOSITION XVI. THEOREM.

The parallelogram formed by drawing tangents through the vertices of two conjugate diameters, is equal to the rectangle of the axes.

Let DED′E′ be a parallelogram, formed by drawing tangents to the conjugate hyperbolas through the vertices of two conjugate diameters DD′, EE′; its area is equal to AA′×BB′.

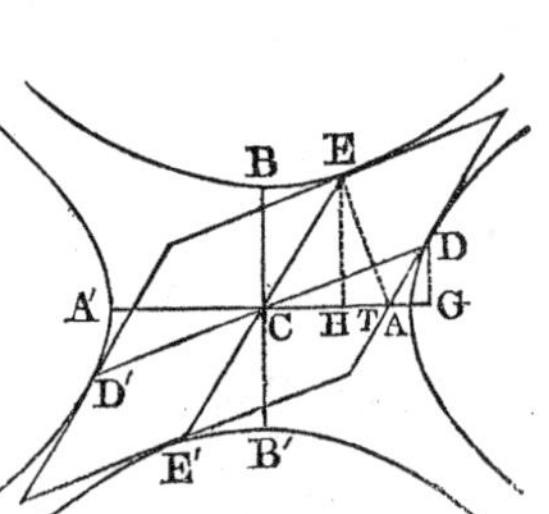

Let the tangent at D meet the major axis in T; join ET, and draw the ordinates DG, EH.

Then, by Prop. XIV., Cor. 2, we have

$$CA^2 : CB^2 :: CG^2 : EH^2,$$

or $\quad CA : CB :: CG : EH.$

But $\quad CT : CA :: CA : CG$ (Prop. X.);

hence $\quad CT : CB :: CA : EH,$

or $\quad CA \times CB$ is equal to $CT \times EH,$

which is equal to twice the triangle CTE, or the parallelogram DE; since the triangle and parallelogram have the same base CE, and are between the same parallels.

Hence $4CA \times CB$ or $AA' \times BB'$ is equal to 4DE, or the parallelogram DED′E′. Therefore, the parallelogram, &c.

PROPOSITION XVII. THEOREM.

If from the vertex of any diameter, straight lines are drawn to the foci, their product is equal to the square of half the conjugate diameter.

Let DD', EE' be two conjugate diameters, and from D let lines be drawn to the foci; then will $FD \times F\,D$ be equal to EC^2.

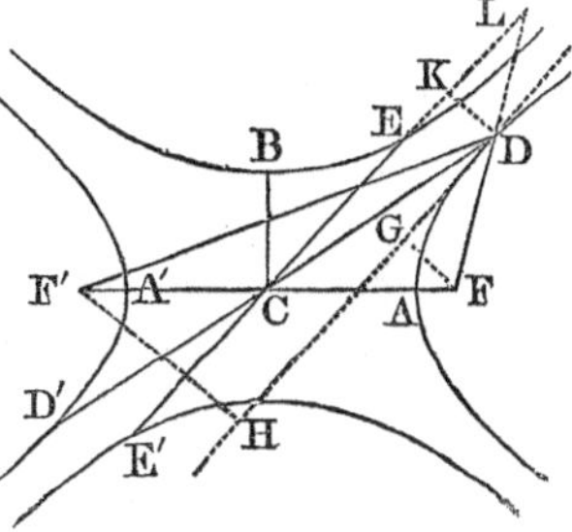

Draw a tangent to the hyperbola at D, and upon it let fall the perpendiculars FG, F'H; draw, also, DK perpendicular to EE'.

Then, because the triangles DFG, DLK, DF'H are similar, we have

$$FD : FG :: DL : DK.$$

Also,
$$F'D : F'H :: DL : DK.$$

Whence (Prop. XI., B. II.),

$$FD \times F'D : FG \times F'H :: DL^2 : DK^2. \qquad (1)$$

But, by Prop. XVI., $AC \times BC = EC \times DK$;

whence
$$AC \text{ or } DL : DK :: EC : BC,$$

and
$$DL^2 : DK^2 :: EC^2 : BC^2. \qquad (2)$$

Comparing proportions (1) and (2), we have

$$FD \times F'D : FG \times F'H :: EC^2 : BC^2.$$

But $FG \times F'H$ is equal to BC^2 (Prop. IX.); hence $FD \times F'D$ is equal to EC^2. Therefore, if from the vertex, &c.

PROPOSITION XVIII. THEOREM.

If a tangent and ordinate be drawn from the same point of an hyperbola to any diameter, half of that diameter will be a mean proportional between the distances of the two intersections from the center.

Let a tangent EG and an ordinate EH be drawn from the same point E of an hyperbola, meeting the diameter CD produced; then we shall have

$$CG : CD :: CD : CH.$$

Produce GE and HE to meet the major axis in K and L; draw DT a tangent to the curve at the point D, and draw DM parallel to GK. Also, draw the ordinates EN, DO.

By Prop. XIV., Cor. 1, $CA^2 : CB^2 :: CO \times OT : DO^2$,

$$:: CN \times NK : EN^2.$$

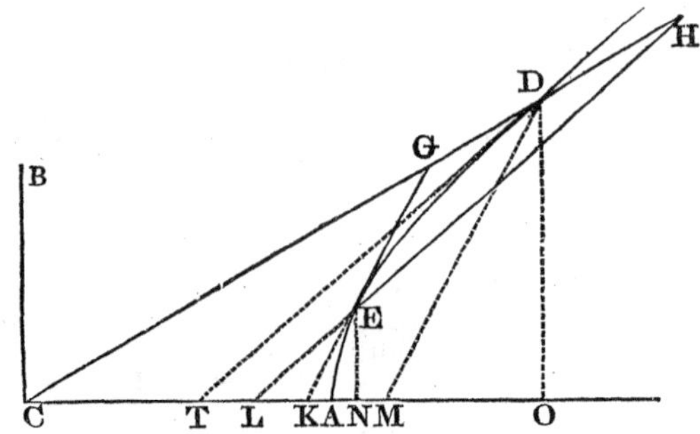

Hence

$CO \times OT : CN \times NK :: DO^2 : EN^2$

$:: OT^2 : NL^2$, by similar triangles. (1)

Also, by similar triangles, $OT : NL :: DO : EN$

$:: OM : NK$. (2)

Multiplying together proportions (1) and (2) (Prop. XI., B. II.), and omitting the factor OT^2 in the antecedents, and $NK \times NL$ in the consequents, we have

$CO : CN :: OM : NL$;

and, by division, $CO : CN :: CM : CL$. (3)

Also, by Prop. X., $CK \times CN = CA^2 = CT \times CO$;

hence $CO : CN :: CK : CT$. (4)

Comparing proportions (3) and (4), we have

$CK : CM :: CT : CL$.

But $CK : CM :: CG : CD$,

and $CT : CL :: CD : CH$;

hence $CG : CD :: CD : CH$.

Therefore, if a tangent, &c.

PROPOSITION XIX. THEOREM.

The square of any diameter, is to the square of its conjugate, as the rectangle of its abscissas, is to the square of their ordinate.

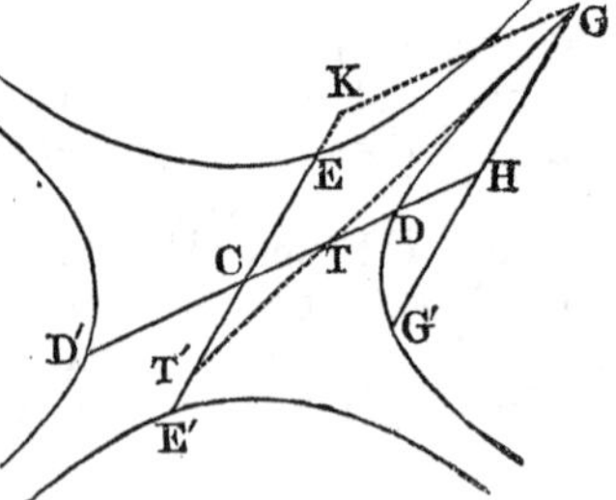

Let DD′, EE′ be two conjugate diameters, and GH an ordinate to DD′; then

$DD'^2 : EE'^2 :: DH \times HD' : GH^2$.

Draw GTT′ a tangent to the curve at the point G, and draw GK an ordinate to EE′. Then, by Prop. XVIII.,

$CT : CD :: CD : CH$

and $CD^2 : CH^2 :: CT : CH$

(Prop. XII., B. II.).

whence, by division, $CD^2 : CH^2 - CD^2 :: CT : HT.$ (1)
Also, by Prop. XVIII., $CT' : CE :: CE : CK$,
and $CE^2 : CK^2 :: CT' : CK$ or GH,
$:: CT : HT.$ (2)

Comparing proportions (1) and (2), we have
$$CD^2 : CE^2 :: CH^2 - CD^2 : CK^2 \text{ or } GH^2,$$
or $$DD'^2 : EE'^2 :: DH \times HD' : GH^2.$$
Therefore, the square, &c.

Cor. 1. In the same manner it may be proved that $DD'^2 : EE'^2 :: DH \times HD' : G'H^2$; hence GH is equal to G′H, or *every diameter bisects its double ordinates.*

Cor. 2. The squares of the ordinates to any diameter, are to each other as the rectangles of their abscissas.

PROPOSITION XX. THEOREM.

If a cone be cut by a plane, not passing through the vertex, and making an angle with the base greater than that made by the side of the cone, the section is an hyperbola.

Let ABC be a cone cut by a plane DGH, not passing through the vertex, and making an angle with the base greater than that made by the side of the cone, the section DHG is an hyperbola.

Let ABC be a section through the axis of the cone, and perpendicular to the plane HDG. Let *bgcd* be a section made by a plane parallel to the base of the cone; then DE, the intersection of the planes HDG, BGCD, will be perpendicular to the plane ABC, and, consequently, to each of the lines BC, HE. So, also, *de* will be perpendicular to *bc* and HE. Let AB and HE be produced to meet in L.

Now, because the triangles LBE, L*be* are similar, as also the triangles HEC, H*ec*, we have the proportions
$$BE : be :: EL : eL$$
$$EC : ec :: HE : He.$$
Hence, by Prop. XI., B. II.,
$$BE \times EC : be \times ec :: HE \times EL : He \times eL.$$
But, since BC is a diameter of the circle BGCD, and DE is perpendicular to BC, we have (Prop. XXII., Cor., B. IV.),
$$BE \times EC = DE^2.$$
For the same reason,
$$be \times ec = de^2.$$

Substituting these values of BE $\times$ EC and $be \times ec$, in the preceding proportion, we have

$$DE^2 : de^2 :: HE \times EL : He \times eL;$$

that is, the squares of the ordinates to the diameter HE, are to each other as the products of the corresponding abscissas. Therefore the curve is an hyperbola (Prop. XII., Cor. 2) whose major axis is LH. Hence the hyperbola is called a *conic section*, as mentioned on page 177.

OF THE ASYMPTOTES.

Definition.—An *asymptote* of an hyperbola is a straight line drawn through the center, which approaches nearer the curve, the further it is produced, but being extended ever so far, can never meet the curve.

PROPOSITION XXI. THEOREM.

If tangents to four conjugate hyperbolas be drawn through the vertices of the axes, the diagonals of the rectangle so formed are asymptotes to the curves.

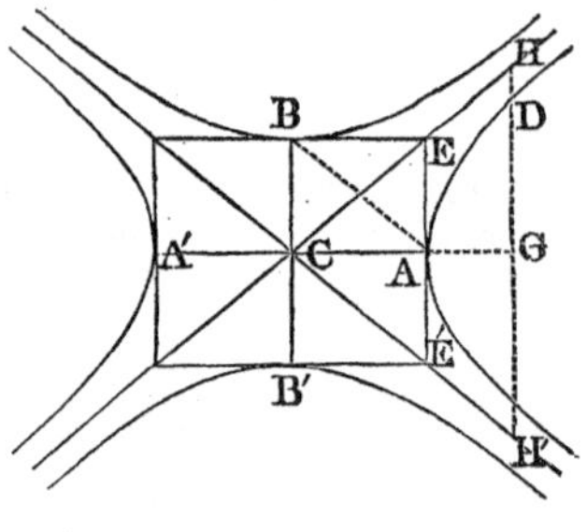

Let AA′, BB′ be the axes of four conjugate hyperbolas, and through the vertices A, A′, B, B′, let tangents to the curve be drawn, and let CE, CE′ be the diagonals of the rectangle thus formed; CE and CE′ will be asymptotes to the curves.

From any point D of one of the curves, draw the ordinate DG, and produce it to meet CE in H. Then, from similar triangles, we shall have

$$CG^2 : GH^2 :: CA^2 : AE^2 \text{ or } CB^2,$$
$$:: CG^2 - CA^2 : DG^2 \text{ (Prop. XII., Cor. 1).}$$

Now, according as the ordinate DG is drawn at a greater distance from the vertex, CG^2 increases in comparison with CA^2; that is, the ratio of CG^2 to $CG^2 - CA^2$ continually approaches to a ratio of equality. But however much CG may be increased, $CG^2 - CA^2$ can never become equal to CG^2; hence DG can never become equal to HG, but approaches continually nearer to an equality with it, the further we recede from the vertex. Hence CH is an asymptote of the hyperbola; since it is a line drawn through the center, which approaches nearer the curve, the further it is produced, but being extended ever so far, can never meet the curve.

In the same manner it may be proved that CH is an asymptote of the conjugate hyperbola.

Cor. 1. The two asymptotes make equal angles with the major axis, and also with the minor axis.

Cor. 2. The line AB joining the vertices of the two axes, is bisected by one asymptote, and is parallel to the other.

Cor. 3. All lines perpendicular to either axis, and terminated by the asymptotes, are bisected by that axis

PROPOSITION XXII. THEOREM.

If an ordinate to either axis be produced to meet the asymptotes, the rectangle of the segments into which it is divided by the curve, will be equal to the square of half the other axis.

Let DG be an ordinate to the major axis, and let it be produced to meet the asymptotes in H and H′; then will the rectangle $HD \times DH'$ be equal to BC^2.

For, by Prop. XII., Cor. 1,

$$CA^2 : AE^2 :: CG^2 - CA^2 : DG^2;$$

or, by similar triangles,

$$:: CG^2 : GH^2.$$

Hence

$$CG^2 : GH^2 :: CG^2 - CA^2 : DG^2,$$

and, by division,

$$CG^2 : GH^2 :: CA^2 : GH^2 - DG^2, \text{ or as } CA^2 : AE^2.$$

Since the antecedents of this proportion are equal to each other, the consequents must be equal; that is,

$$AE^2 \text{ or } BC^2 \text{ is equal to } GH^2 - DG^2;$$

which is equal to $HD \times DH'$.

So, also, it may be proved that

$$CA^2 = D'K \times D'L.$$

Cor. $HD \times DH' = BC^2 = KM \times MK'$; that is, if ordinates to the major axis be produced to meet the asymptotes, the rectangles of the segments into which these lines are divided by the curve, are equal to each other.

PROPOSITION XXIII. THEOREM.

All the parallelograms formed by drawing lines from any point of an hyperbola parallel to the asymptotes, are equal to each other.

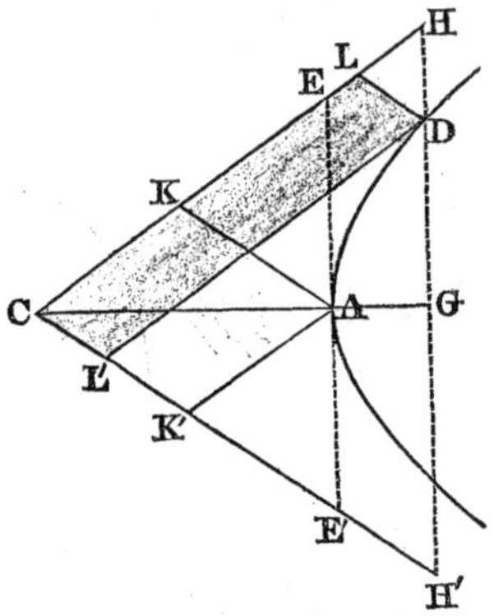

Let CH, CH′ be the asymptotes of an hyperbola; let the lines AK, DL be drawn parallel to CH′, and the lines AK′, DL′ parallel to CH; then will the parallelogram CLDL′ be equal to the parallelogram CKAK′.

Through the points A and D draw EE′, HH′, perpendicular to the major axis; then, because the triangles AEK, DHL are similar, as also the triangles AE′K′, DH′L′, we have the proportions

AK : AE : : DL : DH.

Also, AK′ : AE′ : : DL′ : DH′.

Hence (Prop. XI., B. II.),

AK×AK′ : AE×AE′ : : DL×DL′ : DH×DH′.

But, by Prop. XXII., the consequents of this proportion are equal to each other; hence

AK×AK′ is equal to DL×DL′.

But the parallelograms CA, CD being equiangular, are as the rectangles of the sides which contain the equal angles (Prop XXIII., Cor. 2, B. IV.); hence the parallelogram CD is equal to the parallelogram CA.

Cor. Because the area of the rectangle DL×DL′ is con stant, DL varies inversely as DL′; that is, as DL′ increases, DL diminishes; hence the asymptote continually approaches the curve, but never meets it. The asymptote CH may, therefore, be considered as a tangent to the curve at a point infinitely distant from C.

NOTES.

PAGE 9, *Def.* III.—For the sake of brevity, the word *line* is often used to designate a straight line.

P. 12, *Ax.* II.—This axiom, when applied to geometrical magnitudes, must be understood to refer simply to equality of areas. It is not designed to assert that, when equal triangles are united to equal triangles, the resulting figures will admit of coincidence by superposition.

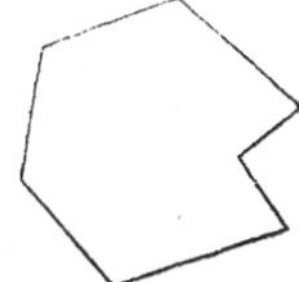

P. 32, Prop. XXVIII.—When this proposition is applied to polygons which have *re-entering* angles, each of these angles is to be regarded as greater than two right angles. But, in order to avoid ambiguity, we shall confine our reasoning to polygons which have only *salient* angles, and which may be called *convex polygons.* Every convex polygon is such, that a straight line, however drawn, can not meet the perimeter of the polygon in more than two points.

P. 32, *Cor.* 2.—This corollary supposes that all the sides of the polygon are produced outward in the *same direction.*

P. 53, Props. XII. and XIII.—It will be perceived that the relative situation of two circles may present five cases.

1st. When the distance between their centers is greater than the sum of their radii, there can be neither contact nor intersection.

2d. When the distance between their centers is equal to the sum of their radii, there is an external contact.

3d. When the distance between their centers is less than the sum of their radii, but greater than their difference, there is an intersection.

4th. When the distance between their centers is equal to the difference of their radii, there is an internal contact.

5th. When the distance between their centers is less than the difference of their radii, there can be neither contact nor intersection.

P. 55, *Cor.* 1.—An angle inscribed in a segment is the angle contained by two straight lines drawn from any point in the circumference of the segment to the extremities of the chord, which is the base of the segment.

P. 63, Prop. VIII.—Every right-angled parallelogram, or *rectangle,* is said to be contained by any two of the straight lines which are about one of the right angles

P. 70, *Scholium.*—By the segments of a line we understand the portions into which the line is divided at a given point. So, also, by the segments of a line produced to a given point, we are to understand the distances between the giv en point and the extremities of the line.

P. 71, Props. XVIII. and XIX.—It will be perceived by these two propositions, that when the angles of one triangle are respectively equal to those of another, the sides of the former are proportional to those of the latter, and conversely; so that either of these conditions is sufficient to determine the similarity of two triangles. This is not true of figures having more than three sides; for with re spect to those of only four sides, or quadrilaterals, we may alter the proportion of the sides without changing the angles, or change the angles without altering the sides; thus, because the angles are equal, it does not follow that the sides are proportional, or the converse. It is evident, for example, that by drawing EF parallel to BC, the angles of the quadrilateral AEFD are equal to those of the quadrilateral ABCD, but the proportion of the sides is different. Also, without changing the four sides AB, BC, CD, DA, we can make the point A approach C, or recede from it, which would change the angles.

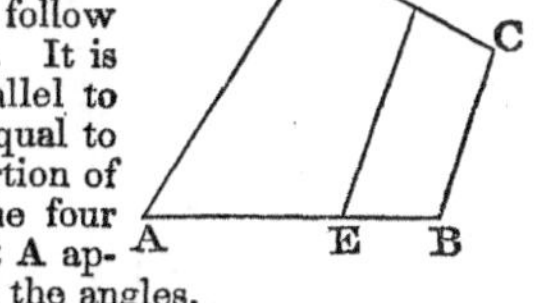

These two propositions, which, properly speaking, form but one, together with Prop. XI., are the most important and the most fruitful in results of any in Geometry. They are almost sufficient of themselves for all subsequent applications, and for the resolution of every problem. The reason is, that all figures may be divided into triangles, and any triangle into two right-angled triangles Thus, the general properties of triangles involve those of all rectilineal figures.

Page 113, Prop. II.—In this and the following propositions, the planes spoken of are supposed to be of indefinite extent.

P. 157, Prop. X.—In all the preceding propositions it has been supposed, in conformity with *Def.* 6, that spherical triangles always have each of their sides less than a semicircumference; in which case their angles are always less than two right angles. For if the side AB is less than a semicircumference, as also AC, both of these arcs must be produced, in order to meet in D. Now the two angles ABC, DBC, taken together, are equal to two right angles; therefore the angle ABC is by itself less than two right angles.

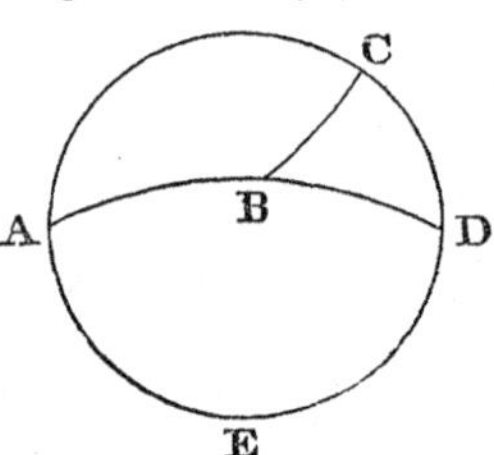

It should, however, be remarked that there are spherical triangles, of which certain sides are greater than a semicircumference, and certain angles greater than two right angles. For if we produce the side AC so as to form an entire circumference, ACDE, the part which remains, after taking from the surface of the hemisphere the triangle ABC, is a new triangle, which may also be designated by ABC, and the sides of which are AB, BC, CDEA. Here we see that the side CDEA is greater than the semicircumference DEA, and at the same time the opposite angle ABC exceeds two right angles by the quantity CBD.

Triangles whose sides and angles are so large have been excluded by the definition, because their solution always reduces itself to that of triangles embraced in the definition. Thus, if we know the sides and angles of the triangle ABC, we shall know immediately the sides and angles of the triangle of the same name, which is the remainder of the surface of the hemisphere.

P. 178.—The *subtangent* is so called because it is below the tangent, being limited by the tangent and ordinate to the point of contact. The *subnormal* is so called because it is below the normal, being limited by the normal and ordinate. The subtangent and subnormal may be regarded as the projections of the tangent and normal upon a diameter.

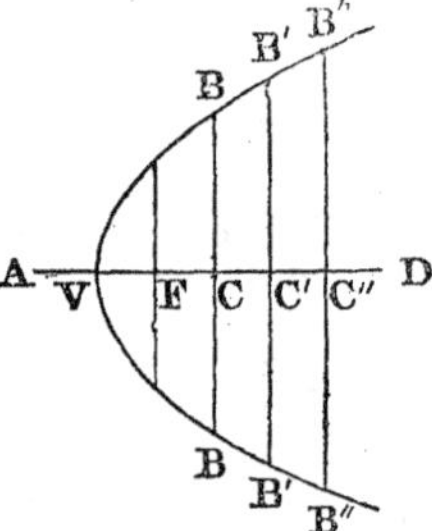

P. 179, Prop. I.—By the method here indicated a parabola may be described with a continuous motion. It may, however, be described by points as follows:

In the axis produced take VA equal to VF, the focal distance, and draw any number of lines, BB, B′B′ etc., perpendicular to the axis AD; then, with the distances AC, AC′, AC″, etc., as radii, and the focus F as a center, describe arcs intersecting the perpendiculars in B, B′, etc. Then, with a steady hand, draw the curve through all the points B, B′, B″, etc.

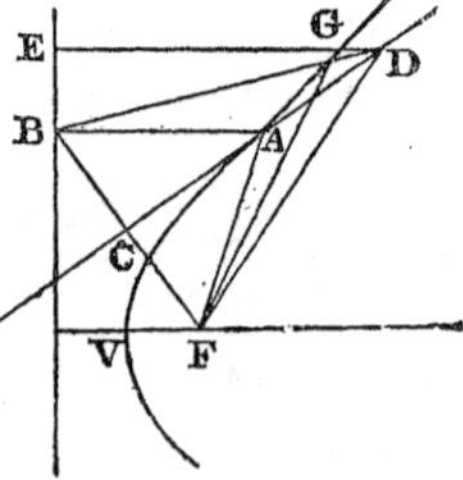

P. 179, Prop. II.—It may be thought that if the point D can not lie *on* the curve, it may fall *within* it, as is represented in the annexed figure. This may be proved to be impossible, as follows:

Let the line DE, perpendicular to the directrix, meet the curve in G, and join FG. Now, by Prop. VIII., B. I.,

$$FG+GD>FD.$$

Hence $FG>FD-GD,$

$$>ED-GD,$$

that is, FG is greater than EG, which is contrary to *Def.* 1.

Page 183, Prop. VIII.—As no attempt is here made to compare figures by superposition, the equality spoken of is only to be understood as implying equal areas. Throughout the remainder of this treatise the word equal is employed instead of equivalent.

P. 185, Prop. XI.—The conclusion that DVG is a parabola would not be legitimate, unless it was proved that the property that "the squares of the ordinates are to each other as the corresponding abscissas" is *peculiar* to the parabola. That such is the case, appears from the fact that, when the axis and one point of a parabola are given, this property will determine the position of every other point. Thus, let VE be the axis of a parabola, and g any point of the curve, from which draw the ordinate ge. Take any other point in the axis, as E, and make GE of such a length that

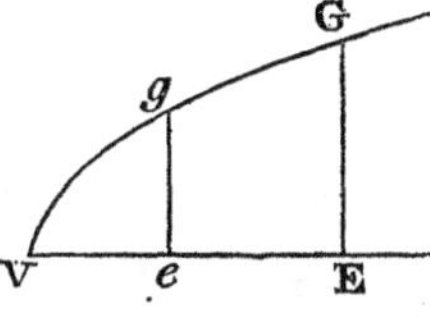

$$Ve : VE :: ge^2 : GE^2.$$

Since the first three terms of this proportion are given, the fourth is determined, and the same proportion will determine any number of points of the curve.

A similar remark is applicable to Prop. XX. of the Ellipse and Hyperbola.

P. 196, Prop. X.—It may be proved that $CT' : CB :: CB : CG'$ in the following manner. Draw DH perpendicular to TT′, and it will bisect the angle FDF′.

Hence

$$F'H : HF :: F'D : DF,$$
$$:: F'T : FT.$$

Therefore, Prop. VII., *Cor.* B. II.,

$$2CF : 2CH :: 2CT : 2CF.$$

Whence $CT \times CH = CF^2$.

But we have proved that

$$CT \times CG = CA^2.$$

Hence $$CT \times GH = CA^2 - CF^2 = CB^2.$$

Again, because the triangles CTT′ and DGH are similar, we have

$$CT : CT' :: DG : GH.$$

Whence $$CT \times GH = CT' \times DG = CT' \times CG';$$

Therefore, $$CT' \times CG' = CB^2,$$

or $$CT' : CB :: CB : CG'.$$

The following demonstration of Prop. X. was suggested to me by Professor J. H. Coffin.

Let TT′ be a tangent to the ellipse, and DG an ordinate to the major axis from

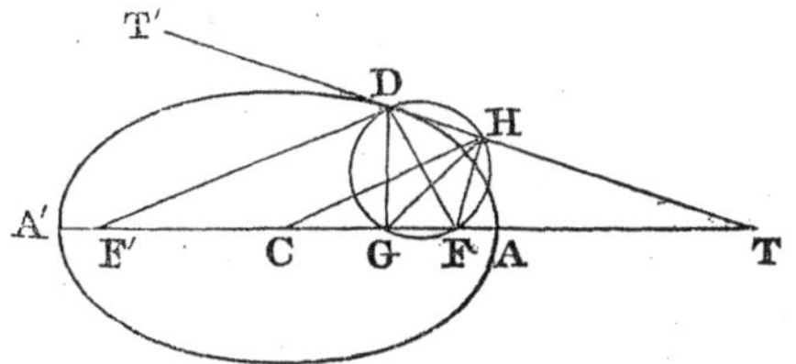

the point of contact; then we shall have

$$CT : CA :: CA : CG.$$

From F draw FH perpendicular to TT′, and join DF, DF′, CH, and GH. Then, by Prop. VIII., *Cor.*, CH is parallel to DF′; and since DGF, DHF are both right angles, a circle described on DF as a diameter will pass through the points G and H. Therefore, the angle HGF is equal to the angle HDF (Prop. XV., *Cor.* 1, B. III), which is equal to T′DF′ or DHC. Hence the angles CGH and CHT which are the supplements of HGF and DHC, are equal. And since the angle C is common to the two triangles CGH, CHT, they are equiangular, and we have

$$CT : CH :: CH : CG.$$

But CH is equal to CA (Prop. VIII); therefore

$$CT : CA :: CA : CG.$$

Page 198, Prop. XIV.—That the triangles CDT, CET′ are similar, may be proved as follows:

$$AG.GA' = CA^2 - CG^2$$
$$= CG.CT - CG^2, \text{ Prop. X.}$$
$$= CG.GT. \qquad (1)$$

In the same manner, AH.HA′=CH.HT′.

Since the triangles DGT, EHC are similar,

$$GT : CH :: DG : EH;$$

or

$$GT^2 : CH^2 :: DG^2 : EH^2;$$
$$:: AG.GA' : AH.HA. \quad \text{Prop. XII., Cor. 2}$$
$$:: CG.GT : CH.HT', \text{ by Equation (1),}$$

Therefore,

$$CG : HT' :: GT : CH$$
$$:: DG : EH.$$

Hence the triangles CDG, EHT′ are similar; and, therefore, the whole triangles CDT, CET′ are similar.

Page 207, Prop. I.—The hyperbola may be described by points, as follows:

In the major axis AA′ produced, take the foci F, F′ and any point D. Then, with the radii AD, A′D, and centers F, F′, describe arcs intersecting each other in E, which will be a point in the curve. In like manner, assuming other points, D′, D″, etc., any number of points of the curve may be found. Then, with a steady hand, draw the curve through all the points E, E′, E″, etc.

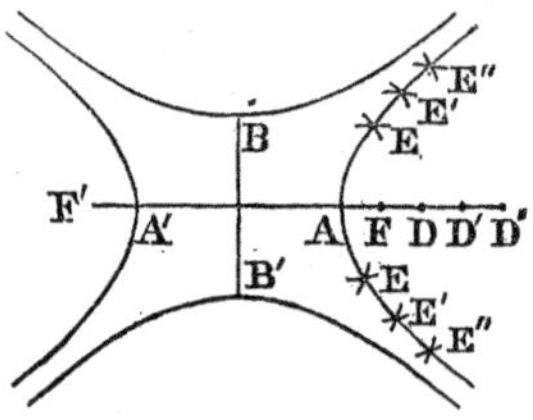

In the same manner may be constructed the two conjugate hyperbolas, employing the axis BB′.

P. 209, Prop. V.—It may be thought that if the point E can not lie *on* the curve, it may fall *within* it, as is represented in the annexed figure. This may be proved to be impossible, as follows:

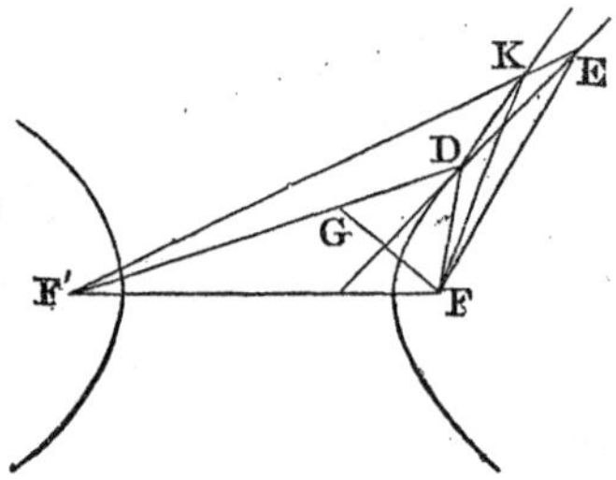

Join EF′, meeting the curve in K, and join KF. Now, by Prop. VIII., B. I.,

$$FK > EF - EK;$$

therefore,

$$F'K - FK < F'K + EK - EF$$
$$< EF' - EF;$$

But $EF' - EF = F'G = DF' - DF$.

Hence $F'K - FK < DF' - DF$,

which is contrary to *Def.* 1.

P. 212, Prop. X.—This proposition may be otherwise demonstrated, like Prop X. of the Ellipse.

GEOMETRICAL EXERCISES.

A FEW theorems without demonstrations, and problems without solutions, are here subjoined for the exercise of the pupil. They will be found admirably adapted to familiarize the beginner with the preceding principles, and to impart dexterity in their application. No general rules can be prescribed which will be found applicable in all cases, and infallibly lead to the demonstration of a proposed theorem, or the solution of a problem. The following directions may prove of some service.

ANALYSIS OF THEOREMS.

1. Construct a diagram as directed in the enunciation, and assume that the theorem is true.
2. Consider what consequences result from this admission, by combining with it theorems which have been already proved, and which are applicable to the diagram.
3. Examine whether any of these consequences are already known to be *true* or to be *false.*
4. If any one of them be false, we have arrived at a *reductio ad absurdum*, which proves that the theorem itself is false, as in Book I., Prop. 4, 16, etc.
5. If none of the consequences so deduced be *known* to be either true or false, proceed to deduce other consequences from all or any of these until a result is obtained which is known to be either true or false.
6. If we thus arrive at some truth which has been previously demonstrated, we then retrace the steps of the investigation pursued in the analysis, till they terminate in the theorem which was assumed. This process will constitute the demonstration of the theorem.

ANALYSIS OF PROBLEMS.

1. Construct the diagram as directed in the enunciation, and suppose the solution of the problem effected.
2. Examine the relations of the lines, angles, triangles, etc., in the diagram, and find the dependence of the assumed solution on some theorem or problem in the Geometry.

3. If such can not be found, draw other lines, parallel or perpendicular, as the case may require; join given points or points assumed in the solution, and describe circles if necessary; and then proceed to trace the dependence of the assumed solution on some theorem or problem in Geometry.

4. If we thus arrive at some previously demonstrated or admitted truth, we shall obtain a direct solution of the problem by assuming the last consequence of the analysis as the first step of the process, and proceeding in a contrary order through the several steps of the analysis, until the process terminate in the problem required.

It may perhaps be expedient to defer attempting the solution of the following problems, until Book V. has been studied.

GEOMETRICAL EXERCISES ON BOOK I.

THEOREMS.

Prop. 1. The difference between any two sides of a triangle is less than the third side.

Prop. 2. The sum of the diagonals of a quadrilateral is less than the sum of any four lines that can be drawn from any point whatever (except the intersection of the diagonals) to the four angles.

Prop. 3. If a straight line which bisects the vertical angle of a triangle also bisects the base, the triangle is isosceles.

Prop. 4. If the base of an isosceles triangle be produced, twice the exterior angle is greater than two right angles by the vertical angle.

Prop. 5. In any right-angled triangle, the middle point of the hypothenuse is equally distant from the three angles.

Prop. 6. If on the sides of a square, at equal distances from the four angles, four points be taken, one on each side, the figure formed by joining those points will also be a square.

Prop. 7. If one angle of a parallelogram be a right angle, the parallelogram will be a rectangle.

Prop. 8. If the diagonals of a quadrilateral bisect each other, the figure is a parallelogram.

Prop. 9. The parallelogram whose diagonals are equal is rectangular.

Prop. 10. Any line drawn through the centre of the diagonal of a parallelogram to meet the sides, is bisected in that point, and also bisects the parallelogram.

PROBLEMS.

Prop. 1. On a given line describe an isosceles triangle, each of whose equal sides shall be double of the base.

Prop. 2. On a given line describe a square, of which the line shall be the diagonal.

Prop. 3. Divide a right angle into three equal angles.

Prop. 4. One of the acute angles of a right-angled triangle is three times as great as the other; trisect the smaller of these.

Prop. 5. Construct an equilateral triangle, having given the length of the perpendicular drawn from one of the angles on the opposite side.

GEOMETRICAL EXERCISES ON BOOK III.

THEOREMS.

Prop. 1. Every chord of a circle is less than the diameter.

Prop. 2. Any two chords of a circle which cut a diameter in the same point, and at equal angles, are equal to each other.

Prop. 3. The straight lines joining toward the same parts, the extremities of any two chords in a circle equally distant from the centre, are parallel to each other.

Prop. 4. The two right lines which join the opposite extremities of two parallel chords, intersect in a point in that diameter which is perpendicular to the chords.

Prop. 5. All the equal chords in a circle may be touched by another circle.

Prop. 6. The lines bisecting at right angles the sides of a triangle, all meet in one point.

Prop. 7. If two opposite sides of a quadrilateral figure inscribed in a circle are equal, the other two sides will be parallel.

Prop. 8. If an arc of a circle be divided into three equal parts by three straight lines drawn from one extremity of the arc, the angle contained by two of the straight lines will be bisected by the third.

Prop. 9. If the diameter of a circle be one of the equal sides of an isosceles triangle, the base will be bisected by the circumference.

Prop. 10. If two circles touch each other externally, and parallel diameters be drawn, the straight line joining the opposite extremities of these diameters will pass through the point of contact.

Prop. 11. The lines which bisect the angles of any parallelogram form a rectangular parallelogram, whose diagonals are parallel to the sides of the former.

Prop. 12. If two opposite sides of a parallelogram be bisected, the lines drawn from the points of bisection to the opposite angles will trisect the diagonal.

PROBLEMS.

Prop. 1. From a given point without a given straight line, draw a line making a given angle with it.

Prop. 2. Through a given point within a circle, draw a chord which shall be bisected in that point.

Prop. 3. Through a given point within a circle, draw the least possible chord.

Prop. 4. Two chords of a circle being given in magnitude and position, describe the circle.

Prop. 5. Describe three equal circles touching one another; and also describe another circle which shall touch them all three.

Prop. 6. How many equal circles can be described around another circle of the same magnitude, touching it and one another?

Prop. 7. With a given radius, describe a circle which shall pass through two given points.

Prop. 8. Describe a circle which shall pass through two given points, and have its centre in a given line.

Prop. 9. In a given circle, inscribe a triangle equiangular to a given triangle.

Prop. 10. From one extremity of a line which can not be produced, draw a line perpendicular to it.

Prop. 11. Divide a circle into two parts such that the angle contained in one segment shall equal twice the angle contained in the other.

Prop. 12. Divide a circle into two segments such that the angle contained in one of them shall be five times the angle contained in the other.

Prop. 13. Describe a circle which shall touch a given circle in a given point, and also touch a given straight line.

Prop. 14. With a given radius, describe a circle which shall pass through a given point and touch a given line.

Prop. 15. With a given radius, describe a circle which shall touch a given line, and have its centre in another given line.

GEOMETRICAL EXERCISES ON BOOK IV.

THEOREMS.

Prop. 1. The area of a triangle is equal to its perimeter multiplied by half the radius of the inscribed circle.

Prop. 2. If from any point in the diagonal of a parallelogram, lines be drawn to the angles, the parallelogram will be divided into two pairs of equal triangles.

Prop. 3. If the sides of any quadrilateral be bisected, and the points of bisection joined, the included figure will be a parallelogram, and equal in area to half the original figure.

Prop. 4. Show how the squares in Prop. XI., Book IV., may be dissected, so that the truth of the proposition may be made to appear by superposition of the parts.

Prop. 5. In the figure to Prop. XI., Book IV.,

(*a.*) If BG and CH be joined, those lines will be parallel.

(*b.*) If perpendiculars be let fall from F and I on BC produced, the parts produced will be equal, and the perpendiculars together will be equal to BC.

(*c.*) Join GH, IE, and FD, and prove that each of the triangles so formed is equivalent to the given triangle ABC.

(*d.*) The sum of the squares of GH, IE, and FD will be equal to six times the square of the hypothenuse.

Prop. 6. The square on the base of an isosceles triangle whose vertical angle is a right angle, is equal to four times the area of the triangle.

Prop. 7. If from one of the acute angles of a right-angled triangle, a straight line be drawn bisecting the opposite side, the square upon that line will be less than the square upon the hypothenuse, by three times the square upon half the line bisected.

Prop. 8. In a right-angled triangle, the square on either of the two sides containing the right angle, is equal to the rectangle contained by the sum and difference of the other sides.

Prop. 9. In any triangle, if a perpendicular be drawn from the vertex to the base, the difference of the squares upon the sides is equal to the difference of the squares upon the segments of the base.

Prop. 10. The squares of the diagonals of any quadrilateral figure are together double the squares of the two lines joining the middle points of the opposite sides.

Prop. 11. If one side of a right-angled triangle is double the other, the perpendicular from the vertex upon the hypothenuse will divide the hypothenuse into parts which are in the ratio of 1 to 4.

Prop. 12. If two circles intersect, the common chord produced will bisect the common tangent.

Prop. 13. The tangents to a circle at the extremities of any chord, contain an angle which is twice the angle contained by the same chord and a diameter drawn from either of the extremities.

Prop. 14. If two circles cut each other, and if from any point in the straight line produced which joins their intersections, two tangents be drawn, one to each circle, they will be equal to one another.

Prop. 15. If from a point without a circle, two tangents be drawn, the straight line which joins the points of contact will be bisected at right angles by a line drawn from the centre to the point without the circle.

PROBLEMS.

Prop. 1. Inscribe a square in a given right-angled isosceles triangle.

Prop. 2. Inscribe a circle in a given rhombus.

Prop. 3. Describe a circle whose circumference shall pass through one angle and touch two sides of a given square.

Prop. 4. In a given square, inscribe an equilateral triangle having its vertex in the middle of a side of the square.

Prop. 5. In a given square, inscribe an equilateral triangle having its vertex in one angle of the square.

Prop. 6. If the sides of a triangle are in the ratio of the numbers 2, 4, and 5, show whether it will be acute-angled or obtuse-angled.

Prop. 7. Given the area and hypothenuse of a right-angled triangle, to construct the triangle.

Prop. 8. Bisect a triangle by a line drawn from a given point in one of the sides.

Prop. 9. To a circle of given radius, draw two tangents which shall contain an angle equal to a given angle.

Prop 10. Construct a triangle, having given one side, the angle opposite to it, and the ratio of the other two sides.

Prop. 11. Construct a triangle, having given the perimeter and the angles of the triangle.

Prop. 12. Upon a given base, describe a right-angled triangle, having given the perpendicular from the right angle upon the hypothenuse.

Prop. 13. Construct a triangle, having given one angle, a side opposite to it, and the sum of the other two sides.

Prop. 14. Construct a triangle, having given one angle, an adjacent side, and the sum of the other two sides

Prop. 15. Trisect a given straight line, and hence divide an equilateral triangle into nine equal parts.

GEOMETRICAL EXERCISES ON BOOK VI.

THEOREMS.

Prop. 1. The square inscribed in a circle is equal to half the square described about the same circle.

Prop. 2. Any number of triangles having the same base and the same vertical angle, may be circumscribed by one circle.

Prop. 3. If an equilateral triangle be inscribed in a circle, each of its sides will cut off one fourth part of the diameter drawn through the opposite angle.

Prop. 4. The circle inscribed in an equilateral triangle has the same centre with the circle described about the same triangle, and the diameter of one is double that of the other.

Prop. 5. If an equilateral triangle be inscribed in a circle, and the arcs cut off by two of its sides be bisected, the line joining the points of bisection will be trisected by the sides.

Prop. 6. The side of an equilateral triangle inscribed in a circle is to the radius, as the square root of three is to unity.

Prop. 7. The sum of the perpendiculars let fall from any point within an equilateral triangle upon the sides, is equal to the perpendicular let fall from one of the angles upon the opposite side.

Prop. 8. If two circles be described, one without and the other within a right-angled triangle, the sum of their diameters will be equal to the sum of the sides containing the right angle.

Prop. 9. If a circle be inscribed in a right-angled triangle, the sum of the two sides containing the right angle will exceed the hypothenuse, by a line equal to the diameter of the inscribed circle.

Prop. 10. The square inscribed in a semicircle is to the square inscribed in the entire circle, as 2 to 5.

Prop. 11. The square inscribed in a semicircle is to the square inscribed in a quadrant of the same circle, as 8 to 5.

Prop. 12. The area of an equilateral triangle inscribed in a circle is equal to half that of the regular hexagon inscribed in the same circle.

Prop. 13. The square of the side of an equilateral triangle inscribed in a circle is triple the square of the side of the regular hexagon inscribed in the same circle.

Prop. 14. The area of a regular hexagon inscribed in a circle is three fourths of the regular hexagon circumscribed about the same circle.

Prop. 15. The triangle, square, and hexagon are the only regular polygons by which the space about a point can be completely filled up.

Prop. 16. The perpendiculars let fall from the three angles of any triangle upon the opposite sides, intersect each other in the same point.

PROBLEMS.

Prop. 1. Trisect a given circle by dividing it into three equal sectors.

Prop. 2. The centre of a circle being given, find two opposite points in the circumference by means of a pair of compasses only.

Prop. 3. Divide a right angle into five equal parts.

Prop. 4. Inscribe a square in a given segment of a circle.

Prop. 5. Having given the difference between the diagonal and side of a square, describe the square.

Prop. 6. Inscribe a square in a given quadrant.

Prop. 7. Inscribe a circle in a given quadrant.

Prop. 8. Describe a circle touching three given straight lines.

Prop. 9. Within a given circle describe six equal circles, touching each other and also the given circle, and show that the interior circle which touches them all, is equal to each of them.

Prop. 10. Within a given circle describe eight equal circles, touching each other and the given circle.

Prop. 11. Inscribe a regular hexagon in a given equilateral triangle.

Prop. 12. Upon a given straight line describe a regular octagon.

THE END.

LOOMIS'S

SERIES OF SCHOOL AND COLLEGE

TEXT-BOOKS.

THE Course of Mathematics by Professor LOOMIS has now been for several years before the Public, and has received the general approbation of Teachers throughout the country. The following are some of the institutions in which this Course has been introduced, either wholly or in part: Dartmouth College, N. H.; Williams College, Mass.; Amherst College, Mass.; Trinity College, Conn.; Wesleyan University, Conn.; Hamilton College, N. Y.; Hobart Free College, N. Y.; New York University, N. Y.; Dickinson College, Penn.; Jefferson College, Penn.; Alleghany College, Penn.; Lafayette College, Penn.; St. James's College, Md.; Emory and Henry College, Va.; Lynchburg College, Va.; Bethany College, Va.; South Carolina, College, S. C.; Alabama University, Ala.; La Grange College, Ala.; Louisiana College, La.; Transylvania University, Ky.; Cumberland College, Ky.; Western Reserve College, Ohio; Marietta College, Ohio; Oberlin College, Ohio; Antioch College, Ohio; Asbury University, Ind.; Wabash College, Ind.; Illinois College, Ill.; Shurtleff College, Ill.; McKendree College, Ill.; Knox College, Ill.; Missouri University, Mo.; ~~University of Michigan, Mich.~~; Beloit College, Wisconsin; Iowa University, Iowa.

Professor Loomis's text-books are distinguished by simplicity, neatness, and accuracy; and are remarkably well adapted for recitation in schools and colleges. I am satisfied no books in use, either in America or England, are so well adapted to the circumstances and wants of American teachers and pupils.—W. C. LARRABEE, *late Professor of Mathematics, Indiana Asbury University.*

Professor Loomis's text-books in Mathematics are models of neatness, precision, and practical adaptation to the wants of students.—*Methodist Quarterly Review.*

These books are terse in style, clear in method, easy of comprehension, and perfectly free from that useless verbiage with which it is too much the fashion to load school-books under pretense of explanation.—*Scott's Weekly Paper, Canada.*

A Treatise on Arithmetic.

Theoretical and Practical. 12mo, 352 pages, Sheep extra, 75 cents.

This volume explains, in a simple and philosophical manner, the theory of all the ordinary operations of Arithmetic, and illustrates them by examples sufficiently numerous to impress them indelibly upon the mind of the pupil. It is designed for the use of advanced students in our public schools, and furnishes a complete preparation for the study of Algebra, as well as for the practical duties of the counting-house.

The answers to about one third of the questions are given in the body of the work; but, in order to lead the student to rely upon his own judgment, the answers to the remaining questions are purposely omitted. For the convenience, however, of such teachers as may desire it, there is published a small edition containing all the answers to the questions.

As an introduction to the author's incomparable series of mathematical works, and displaying, as it does, like characteristic excellences, judicious arrangement, simplicity in the statement, and clearness and directness in the elucidation of principles, this work can not fail of a like flattering reception from the public.—O. L. CASTLE, *Professor of Rhetoric,* and WARREN LEVERETT, A.M., *Principal of Prep. Dep't, Shurtleff College, Illinois.*

We have used Loomis's Arithmetic in this Institute since its publication, and I can truly say that, in arrangement, accuracy, and logical expression it is the best treatise on the subject with which I am acquainted. It has stood the test of the class-room, and I am well pleased with the results.—N. B. WEBSTER, *President of Virginia Collegiate Institute* (Portsmouth).

Professor Loomis has given us a work on Arithmetic which, for precision in language, comprehensiveness of definitions, and suitable explanation, has no equal before the public.—P. E. WILDER, *Greenfield* (Ill.) *Male and Female Seminary.*

I have adopted Professor Loomis's Arithmetic (as well as his entire Mathematical Series) as a text-book in this institution. The rules in this Arithmetic are demonstrated with that unusual clearness and brevity which so pre-eminently distinguish Professor Loomis as a mathematical author.—J. M. FERREE, A.M., *Professor of Mathematics, Dickinson Seminary* (Pa.).

In general arrangement and adaptation to the wants of our schools, I have never seen any thing equal to Professor Loomis's Arithmetic.—DANIEL MCBRIDE, *Bellefonte* (Pa.) *Academy.*

We have taken some pains to examine Professor Loomis's Arithmetic, and find it has claims which are peculiar and pre-eminent. The principles are developed in their natural order; every rule is plainly, though briefly demonstrated, and the pupil is taught to express his ideas clearly and precisely.—*Western Literary Messenger.*

This work is calculated to make scholars thoroughly acquainted with the science of arithmetic. It is certainly superior to any we have ever seen.—*Louisville Courier.*

The clearness and simplicity of Professor Loomis's Arithmetic are in charming contrast with our own reminiscences of similar compilations in our school days, whereof the main and mistaken object was to baffle a child's comprehension.—*The Albion.*

Elements of Algebra.

Designed for the Use of Beginners. Twelfth Edition. 12mo, 281 pages, Sheep extra, 62½ cents.

This volume is intended for the use of students who have just completed the study of Arithmetic. It is believed that it will be found sufficiently clear and simple to be adapted to the wants of a large class of students in our common schools. It explains the method of solving equations of the first degree, with one, two, or more unknown quantities; the principles of involution and of evolution; the solution of equations of the second degree; the principles of ratio and proportion, with arithmetical and geometrical progression. Every principle is illustrated by a copious collection of examples; and two hundred miscellaneous problems will be found at the close of the book.

I have used Loomis's Elements of Algebra in my school for several years, and have found it fitted in a high degree to give the pupil a clear and comprehensive knowledge of the elements of the science. I believe teachers of Academies and High Schools will find it all that they can desire as a text-book on this branch of Mathematics.—Professor ALONZO GRAY, *Brooklyn Heights Seminary.*

I am so much pleased with Loomis's Elements of Algebra that I have introduced it as a text-book in the Institution under my care.—Rev. GORHAM D. ABBOTT, *Spingler Institute, N. Y.*

Loomis's Elements of Algebra is worthy of adoption in our Academies, and will be found to be an excellent text-book. The definitions and rules are expressed in simple and accurate language; the collection of examples subjoined to each rule is sufficiently copious; and as a book for beginners it is admirably adapted to make the learner thoroughly acquainted with the first principles of this important branch of science.—D. MACAULAY, *Principal of the Polytechnic, School, New Orleans.*

Loomis's Algebras form an excellent progressive course for the young student. The "Elements' could be put with advantage into the hands of every child who has mastered the principles of Arithmetic, and is admirably adapted for the use of common schools. The explanations of the author are extremely lucid and comprehensive.—*N. Y. Observer.*

I have carefully examined Loomis's Elements of Algebra, and cheerfully recommend it on account of its superior arrangement and clear and full explanations.—SOLOMON JENNER, *Principal of N. Y. Commercial School.*

Loomis's Elements of Algebra is prepared with the care and judgment that characterize all the elementary works published by the same author.—*Methodist Quarterly Review.*

A Treatise on Algebra.

Eighteenth Edition. 8vo, 359 pages, Sheep extra, $1 00.

This treatise is designed to contain as much of algebra as can be profitably read in the time allotted to this study in most of our colleges, and those subjects have been selected which are most important in a course of mathematical study. Particular pains have been taken to cultivate in the mind of the student a habit of generalization, and to lead him to reduce every principle to its most general form. It is believed that, in respect of difficulty, this treatise need not discourage any youth of fifteen years of age who possesses average abilities, while it is designed to form close habits of reasoning, and cultivate a truly philosophical spirit in more mature minds. In accordance with the expressed wish of many teachers, a classified collection of two hundred and fifty problems is appended to the last edition of this work.

Professor Loomis has here aimed at exhibiting the first principles of Algebra in a form which, while level with the capacity of ordinary students and the present state of the science, is fitted to elicit that degree of effort which educational purposes require. Throughout the work, whenever it can be done with advantage, the practice is followed of generalizing particular examples, or of extending a question proposed relative to a *particular* quantity, to the *class* of quantities to which it belongs, a practice of obvious utility, as accustoming the student to pass from the particular to the general, and as fitted to impress a main distinction between the literal and numerical calculus. The general doctrine of Equations is expounded with clearness and independence. The author has developed this subject in an order of his own. We venture to say that there will be but one opinion respecting the general character of the exposition.—*American Journal of Science and Arts.*

Professor Loomis's Algebra is peculiarly well adapted to the wants of students in academies and colleges. The materials are well selected and well arranged; the rules and principles are stated with clearness and precision, and accompanied with satisfactory proofs, illustrations, and examples.—A. D. STANLEY, *late Professor of Mathematics in Yale College.*

I have carefully examined the work of Professor Loomis on Algebra, and am much pleased with it. The arrangement is sufficiently scientific, yet the order of the topics is obviously, and, I think, judiciously, made with reference to the development of the powers of the pupil. The most rigorous modes of reasoning are designedly avoided in the earlier portions of the work, and deferred till the student is better fitted to appreciate them. All the principles are, however, established with sufficient rigor to give satisfaction. On the whole, therefore, I think this work better suited for the purposes of a text-book than any other I have seen.—AUGUSTUS W. SMITH, LL.D., *President of Wesleyan University.*

Professor Loomis's work is well calculated to impart a clear and correct knowledge of the principles of Algebra. The rules are concise, yet sufficiently comprehensive, containing in few words all that is necessary, and *nothing more;* the absence of which quality mars many a scientific treatise. The collection of problems is peculiarly rich, adapted to impress the most important principles upon the youthful mind, and the student is led gradually and intelligently into the more interesting and higher departments of the science.—JOHN BROCKLESBY, A.M., *Professor of Mathematics and Natural Philosophy in Trinity College.*

I am much pleased with Professor Loomis's Algebra. The arrangement of the subject is, I think, an admirable one. The best proof I can give of the estimation in which I hold it is, that I have taught it to several successive classes in this College.—JOHN TATLOCK, A.M., *Professor of Mathematics in Williams College.*

Professor Loomis's work on Algebra is exceedingly well adapted for the purposes of instruction. He has avoided the difficulties which result from too great conciseness, and aiming at the utmost rigor of demonstration; and, at the same time, has furnished in his book a good and sufficient preparation for the subsequent parts of the mathematical course. I do not know of a treatise which, all things considered, keeps both these objects so steadily in view. —I. WARD ANDREWS, *Professor of Mathematics and Natural Philosophy in Marietta College.*

I regard Professor Loomis's Algebra as altogether worthy of the high reputation its author deservedly enjoys. It possesses those qualities which are chiefly requisite in a college text-book. Its statements are clear and definite; the more important principles are made so prominent as to arrest the pupil's attention; and it conducts the pupil by a sure and easy path to those habits of *generalization* which the teacher of Algebra has so much difficulty in imparting to his pupils.—JULIAN M. STURTEVANT, LL.D., *President of Illinois College.*

Elements of Geometry and Conic Sections.

Fifteenth Edition. 8vo, 234 pages, Sheep extra, 75 cents.

The arrangement of the propositions in this treatise is generally the same as in Legendre's Geometry, but the form of the demonstrations is reduced more nearly to the model of Euclid. The propositions are all enunciated in general terms, with the utmost brevity which is consistent with clearness. The short treatise on Conic Sections appended to this volume is designed particularly for those who have not time or inclination for the study of analytical geometry. The last edition of this work contains a collection of theorems without demonstrations, and problems without solutions, for the exercise of the pupil.

Professor Loomis's Geometry is characterized by the same neatness and elegance which were exhibited in his Algebra. While the logical form of argumentation peculiar to Playfair's Euclid is preserved, more completeness and symmetry is secured by additions in solid and spherical geometry, and by a different arrangement of the propositions. It will be a favorite with those who admire the chaste forms of argumentation of the old school; and it is a question whether these are not the best for the purposes of mental discipline.—*Northern Christian Advocate.*

I consider Loomis's Geometry and Trigonometry the best works that I have ever seen on any branch of elementary mathematics. — JAMES B. DODD, A.M., *Professor of Mathematics, Transylvania University.*

Having used Loomis's Elements of Geometry for several years, carefully examined it, and compared it with Euclid and Legendre, I have found it preferable to either. Teachers will find the work an excellent text-book, suited to give a clear view of the beautiful science of which it treats.—ALONZO GRAY, A.M., *Principal of Brooklyn Heights Seminary.*

Professor Loomis has made many improvements in Legendre's Geometry, retaining all the merits of that author without the defects. I have adopted his work as a text-book in this college.—THOMAS E. SUDLER, A.M., *Professor of Mathematics in Dickinson College.*

Every page of this book bears marks of careful preparation. Only those propositions are selected which are most important in themselves, or which are indispensable in the demonstration of others. The propositions are all enunciated with studied precision and brevity. The demonstrations are complete without being encumbered with verbiage; and, unlike many works we could mention, the diagrams are good representations of the objects intended. We believe this book will take its place among the best elementary works which our country has produced.—*American Review.*

The enunciations in Professor Loomis's Geometry are concise and clear, and the processes neither too brief nor too diffuse. The part treating of solid geometry is undoubtedly superior, in clearness and arrangement, to any other elementary treatise among us.—*N. Y. Evangelist.*

Trigonometry and Tables.

Thirteenth Edition. 8vo, 360 pages, Sheep extra, $1 50. The *Trigonometry* and *Tables* bound separately. The Trigonometry $1 00; Tables, $1 00.

This work contains an exposition of the nature and properties of logarithms; the principles of plane trigonometry; the mensuration of surfaces and solids; the principles of land surveying, with a full description of the instruments employed; the elements of navigation, and of spherical trigonometry. The tables furnish the logarithms of numbers to 10,000, with the proportional parts for a fifth figure in the natural number; logarithmic sines and tangents for every ten seconds of the quadrant, with the proportional parts to single seconds; natural sines and tangents for every minute of the quadrant; a traverse table; a table of meridional parts, &c. The last edition of this work contains a collection of one hundred miscellaneous problems at the close of the volume.

In this work, the principles of Trigonometry and its applications are discussed with the same clearness that characterizes the previous volumes. The portion appropriated to Mensuration, Surveying, &c., will especially commend itself to teachers, by the judgment exhibited in the extent to which they are carried, and the practically useful character of the matter introduced. What I have particularly admired in this, as well as the previous volumes, is the constant recognition of the difficulties, present and prospective, which are likely to embarrass the learner, and the skill and tact with which they are removed. The Logarithmic Tables will be found unsurpassed in practical convenience by any others of the same extent.—AUGUSTUS W. SMITH, LL.D., *President of the Wesleyan University.*

Loomis's Trigonometry is sufficiently extensive for collegiate purposes, and is every where clear and simple in its statements without being redundant. The learner will here find what he really needs without being distracted by what is superfluous or irrelevant.—A. CASWELL, D.D., *Professor of Mathematics and Natural Philosophy in Brown University.*

Loomis's Tables are vastly better than those in common use. The extension of the sines and tangents to ten seconds is a great improvement. The tables of natural sines are indispensable to a good understanding of Trigonometry, and the natural tangents are exceedingly convenient in analytical geometry.—I. WARD ANDREWS, A.M., *Professor of Mathematics and Natural Philosophy in Marietta College.*

Loomis's Trigonometry and Tables are a great acquisition to mathematical schools. I know of no work in which the principles of Trigonometry are so well condensed and so admirably adapted to the course of instruction in the mathematical schools of our country.—THOMAS E. SUDLER, A.M., *Professor of Mathematics in Dickinson College.*

I am so much pleased with Professor Loomis's Trigonometry that I have adopted it as a text-book in this college.—JOHN BROOKLESBY, A.M., *Professor of Mathematics in Trinity College.*

Loomis's Trigonometry is well adapted to give the student that distinct knowledge of the principles of the science so important in the further prosecution of the study of mathematics. The description and representation of the instruments used in surveying, leveling, &c., are sufficient to prepare the student to make a practical application of the principles he has learned. The Tables are just the thing for college students.—JOHN TATLOCK, A.M., *Professor of Mathematics in Williams College.*

Elements of Analytical Geometry,

and of the Differential and Integral Calculus. Eleventh Edition. 8vo, 278 pages, Sheep extra, $1 50.

The first part of this volume treats of the application of algebra to geometry, the construction of equations, the properties of a straight line, a circle, parabola, ellipse, and hyperbola; the classification of algebraic curves, and the more important transcendental curves. The second part treats of the differentiation of algebraic functions, of Maclaurin's and Taylor's Theorems, of maxima and minima, transcendental functions, theory of curves, and evolutes. The third part exhibits the method of obtaining the integrals of a great variety of differentials, and their application to the rectification and quadrature of curves, and the cubature of solids. All the principles are illustrated by an extensive collection of examples, and a classified collection of a hundred and fifty problems will be found at the close of the volume. The work was prepared to meet the wants of the mass of college students of average abilities.

Analytical Geometry is treated, amply enough for elementary instruction, in the short compass of 112 pages, so that nothing may be omitted, and the student can master his text-book as a whole. The Calculus is treated in like manner in 167 pages, and the opening chapter makes the nature of the art as clear as it can possibly be made. We recommend this work, without reserve or limitation, as the best text-book on the subject we have yet seen.—*Methodist Quarterly Review.*

Loomis's Analytical Geometry and Calculus is the best work on that subject for a college course and mathematical schools. It contains all the important principles and doctrines of the calculus, simplified and illustrated by well selected problems.—THOMAS E. SUDLER, A.M., *Professor of Mathematics in Dickinson College.*

Loomis's Calculus is better adapted to the capacities of young men than any book heretofore published on this subject.—A. P. HOOKE, *Professor of Mathematics in Bethany College.*

I have examined Loomis's Analytical Geometry and Calculus with great satisfaction, and shall make it an indispensable part of our scientific course.—JAMES B. DODD, A.M., *Professor of Mathematics in Transylvania University.*

I am well pleased with Loomis's Analytical Geometry and Calculus, as it brings the subjects within the powers of the majority of our students, a thing certainly that very few authors on the Calculus try to do.—JAMES CURLEY, *Professor of Mathematics in Georgetown College.*

No similar work is at the same time so concise and so comprehensive; so well adapted for a college class, wherein every part can be taught in the time prescribed for this department.—I. TOWLER, *Professor of Mathematics in Hobart Free College.*

Introduction to Practical Astronomy.

With a Collection of Astronomical Tables. 8vo, 497 pages, Sheep extra, $1 50.

This work furnishes a description of the instruments required in the outfit of an observatory, as also the methods of employing them, and the computations growing out of their use. It treats particularly of the Transit Instrument and of Graduated Circles; of the method of determining time, latitude, and longitude; with the computation of eclipses and occultations. The work is designed for the use of amateur observers, practical surveyors, and engineers, as well as students who are engaged in a course of training in our colleges. The tables which accompany this volume are such as have been found most useful in astronomical computations, and to them has been added a catalogue of 1500 stars, with the constants required for reducing the mean to the apparent places.

Letters commendatory of this work have been received from G. B. Airy, Astronomer Royal of England; from William Whewell, D.D., Master of Trinity College, Cambridge, England; from Professor J. Challis, Plumian Professor of Astronomy in the University of Cambridge, England; from J. C. Adams, late President of the Royal Astronomical Society; from Augustus De Morgan, Professor of Mathematics in University College, London; from M. J. Johnson, Director of the Radcliffe Observatory, Oxford, England; from William Lassell, Astronomer of Liverpool, England; from C. Piazzi Smyth, Astronomer Royal for Scotland; from the Earl of Rosse, Ireland; from Edward J. Cooper, of Markree Castle Observatory, Ireland; and from numerous astronomers from every part of the United States.

Professor Loomis's work on Practical Astronomy is likely to be extensively useful, as containing the most recent information on the subject, and giving the information in such a manner as to make it accessible to a large class of readers. I am of opinion that Practical Astronomy is a good *educational* subject even for those who may never take observations, and that a work like this of Professor Loomis should be a text-book in every university. The want of such a work has long been felt here, and if my astronomical duties had permitted, I should have made an attempt to supply it. It is remarkable that in England, where Practical Astronomy is so much attended to, no book has been written which is at all adapted to making a learner acquainted with the recent improvements and actual state of the science.—JAMES CHALLIS, *Plumian Professor of Astronomy in the University of Cambridge, England.*

Professor Loomis's volume on Practical Astronomy is by far the best work of the kind at present existing in the English language.—J. P. NICHOL, LL.D, *Professor of Practical Astronomy in the University of Glasgow, Scotland.*

The science of the age was most assuredly in want of a work on Practical Astronomy, and I am delighted to find that want now supplied from America, and from the pen of Professor Loomis. I propose to make this volume a text-book for my class of Practical Astronomy in the University of Edinburgh.—C. PIAZZI SMYTH, *Astronomer Royal for Scotland.*

No work since that of Professor Woodhouse places the reader so directly in communication with the interior of the Observatory as the work on Practical Astronomy by Professor Loomis; and he has supplied a want which young astronomers, actually wishing to observe, must have felt for a long time. It is more than possible that this work may establish itself as a text-book in England. —AUGUSTUS DE MORGAN, *Professor of Mathematics in University College, London.*

Recent Progress of Astronomy,

especially in the United States. Revised Edition. 12mo, 396 pages, Muslin, $1 00.

This volume is designed to exhibit, in a popular form, the most important astronomical discoveries of the past ten years. It treats particularly of the discovery of the planet Neptune, of the new asteroids, of the new satellite, and the new ring of Saturn, of the great comet of 1843, Biela's comet, Miss Mitchell's comet, &c.; of the parallax of fixed stars, motion of the stars, resolution of the nebulæ, &c.; the history of American observatories, determination of longitude by the electric telegraph, manufacture of telescopes in the United States, &c. The new edition of this work has been mostly re-written and much enlarged, and contains the most important discoveries in Astronomy down to the present time.

Professor Loomis's view of the circumstances attending the discovery of Neptune appears to me the truest and most impartial that I have seen. —J. CHALLIS, *Plumian Professor of Astronomy in the University of Cambridge, England.*

I thank you for your interesting little work on the Recent Progress of Astronomy: you have reason to be proud of the rapid advances which science in general, and especially Astronomy, has lately made in America.—J. C. ADAMS, *late President of the Royal Astronomical Society.*

Professor Loomis's volume on the Recent Progress of Astronomy contains a great deal of useful and valuable information. What is said about American observatories was in great part new to me.—J. LAMONT, *Director of the Astronomical Observatory, Munich, Bavaria.*

The design of this work is to exhibit, in a popular form, the most important astronomical discoveries of the last ten years. The author has executed the task with his usual thoroughness and accuracy, and the student is here furnished, in a condensed and reliable form, with a large amount of important information, to collect which from the original sources would cost him much time and labor.—*American Journal of Science and Arts.*

Loomis's "Recent Progress of Astronomy" has afforded me great interest, for it is admirably done. As a work to be read by a multitude of our intelligent people who are not adepts in astronomy, it has no competitor. It supplies a desideratum that was strongly felt, and must gratify numbers who are interested in the progress of astronomy in our own country.—CHESTER DEWEY, LL.D., *Professor in Rochester University.*

Elements of Natural Philosophy and Astronomy,

for the Use of Academies and High Schools. 12mo, Sheep. (*In press.*)

This volume exhibits in a concise form the fundamental principles of Natural Philosophy and Astronomy, arranged in their natural order, and explained in a clear and scientific manner, without requiring a knowledge of the mathematics beyond that of the elementary branches. Special pains have been taken to make this work both practical and interesting by borrowing illustrations from common life, and by explaining phenomena which are familiar to all, but whose philosophy is not generally well understood.

Published by HARPER & BROTHERS,

Franklin Square, New York.

HARPER & BROTHERS will send either of the above Works by Mail, postage paid (for any distance in the United States under 3000 miles), on receipt of the Money.

Harper's Catalogue.

A New Descriptive Catalogue of Harper & Brothers' Publications, with an Index and Classified Table of Contents, is now ready for Distribution, and may be obtained gratuitously on application to the Publishers personally, or by letter inclosing Six Cents in Postage Stamps.

The attention of gentlemen, in town or country, designing to form Libraries or enrich their Literary Collections, is respectfully invited to this Catalogue, which will be found to comprise a large proportion of the standard and most esteemed works in English Literature —COMPREHENDING MORE THAN TWO THOUSAND VOLUMES — which are offered, in most instances, at less than one half the cost of similar productions in England.

To Librarians and others connected with Colleges, Schools, &c., who may not have access to a reliable guide in forming the true estimate of literary productions, it is believed this Catalogue will prove especially valuable as a manual of reference.

To prevent disappointment, it is suggested that, whenever books can not be obtained through any bookseller or local agent, applications with remittance should be addressed direct to the Publishers, which will be promptly attended to.

www.ingramcontent.com/pod-product-compliance
Lightning Source LLC
LaVergne TN
LVHW050522100826
845148LV00002B/416

* 9 7 8 1 4 2 5 5 2 0 8 2 3 *